The Beekeeper's Manual

The Beekeeper's Manual

L. A. Stephens-Potter

DAVID & CHARLES
Newton Abbot   London   North Pomfret (Vt)

To my mother, who waited so long for me to achieve something

British Library Cataloguing in Publication Data

Stephens-Potter, L. A.
  The beekeeper's manual
  1. Bee culture
  I. Title
  638′.1   SF523

  ISBN 0–7153–8485–6

Photoset in Monophoto Plantin
by Latimer Trend & Co Ltd, Plymouth
and printed in Great Britain
by Redwood Burn Ltd, Trowbridge, Wilts
for David & Charles (Publishers) Limited
Brunel House Newton Abbot Devon

Published in the United States of America
by David & Charles Inc
North Pomfret Vermont 05053 USA

Contents

Introduction

When I decided to attempt keeping bees and looked to the local library for information, I found that there were a number of publications that dealt with beekeeping but that most were very detailed and therefore confusing to the absolute beginner. A few were charming but romantic nonsense; others presented a rigid dogma and claimed that only strict adherence to the methods and equipment specified could possibly yield profitable results.

This book is designed to be an easy-to-follow reference manual for the beginner. It is not intended to be a detailed or comprehensive work which delves into the more specialised or obscure aspects of bee lore, but should be used rather as a simple primer. It is presented with an economy of words and without prejudice. All advice or comments, while based on fact, are intended for guidance only, and readers should remember that it is the welfare of the bees that is of prime importance in beekeeping; the production of honey is an attractive but secondary consideration.

1 Beekeeping as a Hobby

Who keeps bees?

Most of the honey bees in this country are kept by amateurs as a part-time hobby, often with only a few hives. Except for those unfortunate people who are allergic to bee stings, anybody with sufficient interest and an enquiring mind is capable of learning the basic skills for the good husbandry of bees. Age or sex is of no consequence as long as the individual is in reasonable health and enjoys an outdoor pursuit.

At present, there seems to be a general revival of interest in beekeeping. Only a few years ago the description 'beekeeper' would have conjured up visions of a mildly eccentric, kindly old man with a cottage-garden full of neat white-painted hives. Of course, no occupation is so stereotyped, and people from all walks of life indulge in this fascinating hobby. Even an urban dweller with a modest garden can, with careful siting of the hives, keep a stock or two.

Half-truths and superstitions

Beekeeping seems to attract superstitions, and even in these latter years of the twentieth century it retains an air of mystery. Admittedly, there are always individuals who use the ignorance of others to project the image that keeping bees is a form of 'white magic', and therefore to be a beekeeper is to belong to an exclusive club. I am not sure whether this is intended as a form of one-up-manship or a misguided attempt to keep the production of honey limited and the market price buoyant. As it is, superstitions linger, although it is doubtful whether many people still feel the need to inform bees when their owner has died. According to tradition, if this ritual were not observed and the hives preferably decorated with black crêpe, the bees would die or depart from the hive. It is,

of course, quite conceivable that some stocks of bees died out or swarmed after their owner passed away because they were neglected! The truth is often much less colourful than fiction.

The following half-truths and super-stitions connected with beekeeping are still bandied about but can, I think, be viewed as fairly harmless nonsense:

A savage stock of bees always gives more honey.

In fact, because such bees tend to be excitable they probably swarm more frequently and their true work potential is never realised. There is also little pleasure to be derived from dealing with such vicious insects.

The little black bees are always more spiteful.

This is not necessarily so. It would be just as erroneous to declare that the temperament of a mongrel dog could be determined by its colour or the shape of its ears.

I don't like flat roofs on hives: the water won't run off.

Flat roofs will not shed water in such a uniform manner as a gabled roof, but if water piles up on a flat roof then it has not only escaped my observation but has also defied gravity!

I could never be a beekeeper—bees don't like me.

As far as I am aware, bees have no preference as to who handles them, but do respond to how they are handled. In other words, the fault does not lie with the bees.

Inevitably, a number of superstitions concern swarming or, more specifically, a swarm in flight. 'Tanging' used to be

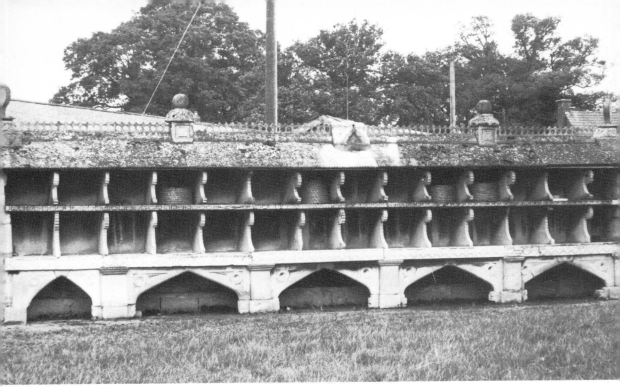

A bee bole. Prior to removable frame hives, skeps were often kept in this form of construction. This is a particularly magnificent example of its type *(by kind permission of Gloucestershire College of Agriculture)*

performed when a swarm appeared which was supposed to aid recovery of the swarm by causing it to alight. It was carried out by striking a metal object, such as a frying pan. The most suitable 'striker' was traditionally a key and the metallic din produced was believed to keep the swarm in check. Throwing clods of earth or handfuls of soil into the swarm in flight was supposed to have the same effect, but many people with their understandable, though unreasonable, fear of swarms would probably be reluctant to try the latter method.

Time with the bees

Most forms of livestock require regular, often daily, attention. This is not so with bees. The time spent with them is generally proportional to the number of colonies you keep and the time of year. From April to September it may be necessary to attend to the bees about once a week, but only occasionally during the remainder of the year. As bees do not require daily attention it is possible to site them some distance from home and still maintain the necessary care. If possible, though, keep them in the garden; there is much to be learned from close observation.

One word of warning: time passes so quickly when absorbed with bees that it is very easy to be late for meals or for appointments. I used to keep six colonies in an orchard behind the house and if I started work with them at about 10.30am I would sometimes be hard pressed to be home for Sunday lunch at 1.00pm. Examination of the six colonies could be completed in about one and a half hours, but once I became absorbed in studying some small detail it was not long before the church clock was striking midday and I sometimes had four colonies still to attend to.

Beekeeping organisations

There are a number of bodies concerned with beekeeping organised from local to national level, most being affiliated to or directly linked with the British Beekeepers Association. These include the following: the Federation of Irish Beekeepers, the

Scottish Beekeepers Association, the Welsh Beekeepers Association, the Ulster Beekeepers Association, the National Honey Show, the Bee Farmers Association, the Gloucester Honey Bee Improvement Group, the Apiculture Education Association, the British Isles Bee Breeders Association and the International Bee Research Association. Details of most of these organisations can be found in an excellent publication, *The Beekeeper's Annual*, published by Northern Bee Books, Scout Bottom Farm, Mytholmroyd, Hebden Bridge, West Yorkshire, HX7 5JS.

There are several advantages of belonging to an organisation, especially on a local level. Not only can you meet other beekeepers, many of whom will have been keeping bees for many years and have a vast fund of experience to call on and many amusing anecdotes to relate, but there are also sales of secondhand equipment, which are often useful to get the beginner started. Insurance

can be arranged against the loss of bees and equipment by damage or disease, and specialist knowledge is often available at a county level where the specialist involved may also be the beekeeping instructor at the local college.

For the more studious, there is an Education and Examination Board which covers different levels and aspects of beekeeping. Details of the British Beekeepers Association and the name and address of your local branch secretary may be obtained from: The Secretary, British Beekeepers Association, National Beekeeping Centre, Stoneleigh, Warwickshire, CV8 2LZ. Specialist advice on bee diseases (and where adult bees may be sent for examination) may be obtained from: The National Beekeeping Unit, Luddington Experimental Horticulture Station, Stratford upon Avon, CV37 9SJ.

Some basic terms

We all know that bees are kept in hives, but it becomes a little more complicated when one hears of WBCs, Nationals and box hives. They are, of course, all hives of different designs made to contain bees and exercise control over their activities. There are also

The WBC hive, regarded by many beekeepers as a 'classic'. A warm, dry hive for the winter, ideal for a garden setting. This exhibition hive is fitted with five 'lifts'; normally three would be used. Note the gabled roof and flight board *(by kind permission of the British Beekeepers Association)*

British National hives in E. H. Thorne's apiary *(E. H. Thorne (Beehives) Ltd)*

skeps, primitive hives made from straw and kept in bee boles which basically comprise a system of shelves built into a wall to hold skeps.

Of the bees, there are queens (usually one per hive), drones (male bees which cannot sting) and workers (imperfect female bees which can, and do, sting). Bees swarm to reproduce their own kind by splitting the colony. The kind of expression 'bees swarming all over a bush' is not strictly correct as it is not a swarming function, but this use of the word has grown into the English language.

Bees kept in hives may be referred to as 'colonies' or 'stocks'. I choose to refer to them as so many colonies or hives of bees, but another beekeeper may say that he or she has some six 'stocks' of bees. 'Supering' is the name given to the activity of adding extra containers to the hive for the bees to store surplus honey. The 'supers' are taken away to be cut up for comb honey or to have their honey extracted and bottled.

A whole range of equipment is available to aid the keeping of bees and to make use of their produce. Many of these devices bear the name of their designer or manufacturer, for instance, there are Porter bee escapes, MG wax extractors, Smith and Langstroth hives and Waldron excluders. The terminology associated with beekeeping is interesting, and I would advise anyone just starting to obtain a manufacturer's catalogue and study it closely. To the uninitiated, it is quite fascinating.

2 The Bees

The colony

Bees are not individuals; they live with many thousands of other bees in a colony and respond to certain instincts and stimuli. Bees have many abilities which are difficult for a human to understand; for instance, they can navigate to and from a hive that has been moved by it's owner several miles overnight to a new site, and they conduct all the business of the hive—breeding, storing pollen and honey—in complete darkness; only when the hive is opened for inspection does light normally enter. The honey bee is a highly efficient insect which has been perfected for millions of years. There are slight variations in colour, temperament and breeding strains, but as an insect type the honey bee seems perfectly adapted for its task.

In every colony of bees there are workers, drones and a queen. The workers number many thousands, while the drones, which only live during the summer months, are normally only counted in their hundreds. For most of the year there is only one queen, but others can be produced and are produced to enable the colony to continue and multiply. This multiplication is effected by the act of swarming and can be likened to an amoeba splitting, as one queen and several thousand other bees detach themselves from the parent colony to form another colony at a new location.

The brood nest

In common with most insects, the honey bee begins life as an egg. These eggs are laid by

The start of a queen cell. These may sometimes only be incomplete so-called 'play' cells, and will not be built any larger than the examples shown *(author)*

Capped brood and grubs. Note the bee emerging from its cell (*author*)

the queen in cells of differing size and construction, depending on whether the final product will be a queen, drone or worker. These masses of cells are what we recognise as the wax comb, and in a hive or wild stock the nearer the centre of the brood nest the greater the area of the comb used for breeding. Around this brood area is a stock of pollen and that is ringed by the stored honey. Although it is usual and convenient to use rectangular frames for the comb in modern hives, they do not conform to the natural shape of an expanding brood nest. This does not seem to matter much in practice, however, as the corners of the comb in a hive are usually crammed with stored honey.

The cells in which worker and drone bees are bred are hexagonal when viewed at the open end, approximately ⅜in (10mm) deep with a concave base. This is the ideal shape both for strength and for providing the maximum number of cells for a given area. A piece of comb has two sides to it with the cells arranged on both sides. These cells are not at right angles to an imaginary centre line drawn through the comb but are inclined with the base lower than the mouth of the cell. Although worker and drone cells are the same in shape and type of construction they differ in size, the drone cells being larger.

The queen cells are quite different from the worker and drone cells which are also used for storing food. The base of the queen cell is attached to the comb but the body of it is then drawn out until it hangs down almost parallel to the parent comb. It is rather like an extended acorn in appearance, and is round inside instead of hexagonal; it is also the largest type of cell in a hive. Queen cells are usually few in number and may be found almost anywhere on the comb. During most seasons the bees will start to build queen cells which are never completed; these are often referred to as 'play cells'.

The breeding cycle

The honey bee egg hatches after three days to become a larva or grub. This three-day hatching period is the same for queen, drone and worker; after that, although the process of change is the same for all three, they differ in their rate of growth.

The larva is fed by young bees, so-called 'nursery bees', and grows rapidly until it almost fills the cell. The cell is then sealed

14

with a porous capping which is slightly domed and quite different in appearance from the wax capping which covers the stored honey; this is known as sealed brood. Inside the sealed cells the larva spins a cocoon and changes to a pupa.

Worker bee cells are sealed for twelve days at the end of which a perfect worker bee emerges. During those twelve days the larva undergoes that remarkable transformation called metamorphosis, changing from soft white grub to a cocooned pupa and finally a winged insect capable of performing a remarkable number of functions. The development from egg to insect for queen, worker and drone is shown in Fig 1 (page 16).

Breeding is not continued at a constant rate throughout the year. There is a period of comparative inactivity during the winter months when breeding does not take place. From about February onwards the queen starts laying, reaching a peak in July. After this, the rate of egg laying diminishes, although well into August most hives will appear to be overflowing with bees if they are healthy and have a strong laying queen.

Bees on the comb. The queen bee is indicated in the centre—note the length compared with the worker bees (*author*)

Anatomy of a bee (Fig 2)

The body of a honey bee is divided into three main parts, the head, thorax and abdomen. The hard outside segments of the body form the exoskeleton; bees have the skeleton on the outside with muscles located internally.

On the head are the eyes, mouth and a pair of antennae or feelers. The compound eyes of the drone are particularly large and give the impression that they almost cover the head. The mouth is located at the lower front of the head between a pair of mandibles or jaws, and the proboscis, a sort of tongue, is used to draw up nectar for feeding or storing in the 'honey stomach' in the abdomen. The antennae have sensory endings which transmit to the brain, within the head, information concerning, among other things, scent and taste.

The thorax or middle section of the body carries two pairs of wings and six legs externally, and houses the locomotive muscles and the trachea or breathing tubes internally. The thorax is joined to the abdomen by a narrow 'neck' which always gives insects a curious disjointed appearance. The abdomen is composed of six telescopic

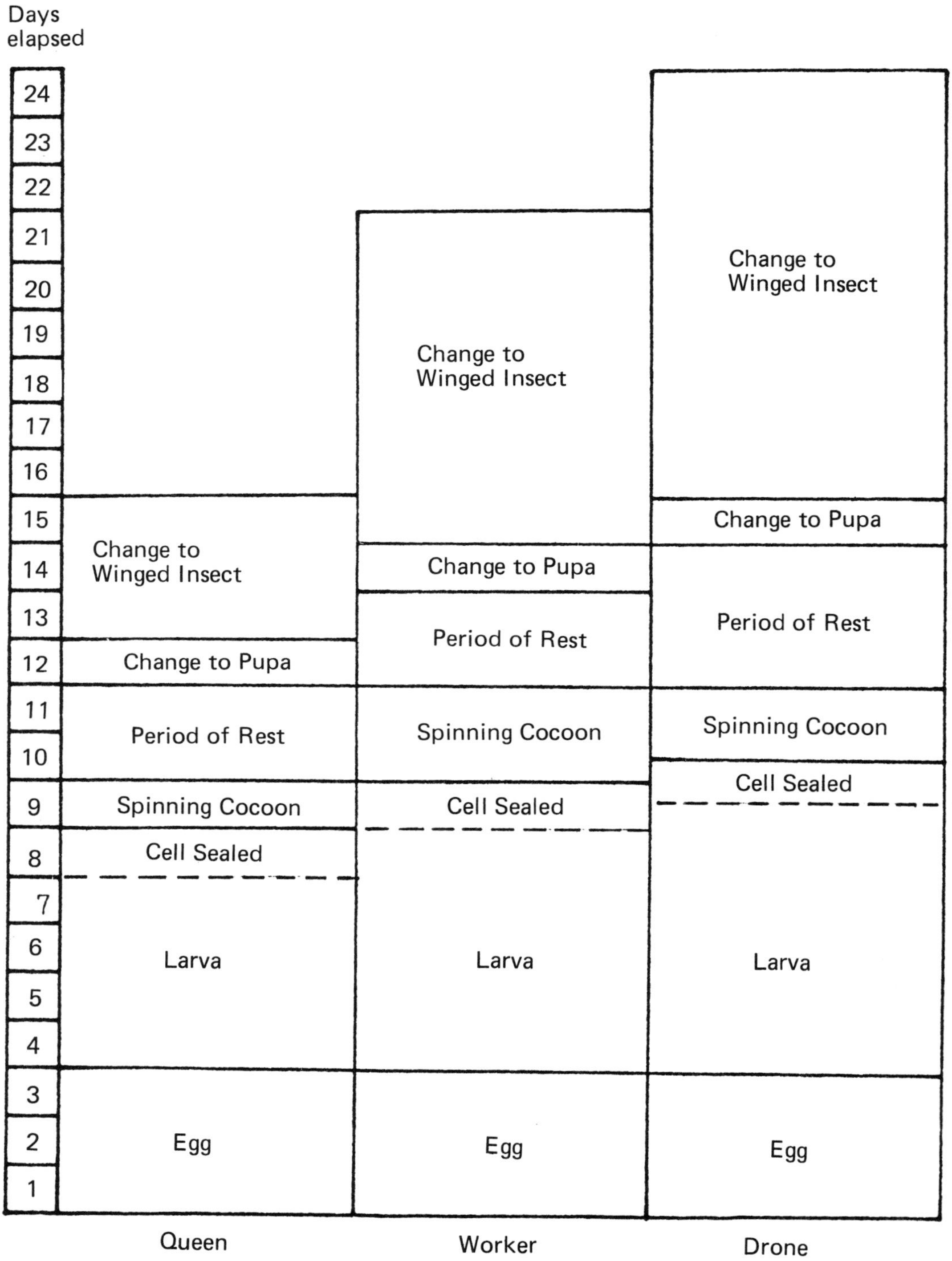

Days
elapsed

	Queen	Worker	Drone
24			Change to Winged Insect
23			
22			
21		Change to Winged Insect	
20			
19			
18			
17			
16			
15	Change to Winged Insect		Change to Pupa
14		Change to Pupa	
13		Period of Rest	Period of Rest
12	Change to Pupa		
11	Period of Rest	Spinning Cocoon	Spinning Cocoon
10			
9	Spinning Cocoon	Cell Sealed	Cell Sealed
8	Cell Sealed		
7	Larva	Larva	Larva
6			
5			
4			
3	Egg	Egg	Egg
2			
1			

Fig 1   The development of a bee

segments within which are located the 'honey stomach', intestinal passages, heart, wax glands, scent glands, reproductive organs and, at the tail end, the sting, except for the drone which does not possess a sting.

This is only a very simple description of a remarkably complex insect, but most of us amateurs are initially more concerned with the activities and performance of our bees than with detailed anatomical descriptions.

The queen

The queen is the most important member of the hive and is, at the same time, both mother and servant of all. She is not the ruler or leader of the colony, and is probably the least versatile but most specialised member. Her sole function is to lay eggs, thereby ensuring the continuation of the colony.

To human eyes the queen has a graceful, well-proportioned appearance. Her body is longer than that of the worker, her wings proportionally shorter and she has a long, tapered abdomen. A queen possesses a sting but uses it only to attack other queens in the hive which she quite rightly regards as rivals. Consequently, there is normally only one queen present in a colony.

The queen is developed from a fertilised egg that is identical to those eggs from which worker bees form. The difference in their development is brought about by cell size and a special diet. Larvae destined to become queens are fed a special food in unrestricted quantities which has become known as 'royal jelly'. It is produced by worker bees of a certain age as is the common brood food supplied to all newly hatched larvae. It is believed that this 'royal jelly' contains special substances so that at the end of sixteen days after the laying of an egg a perfect female emerges. This female can, after mating with a drone, produce sufficient eggs to maintain a colony of bees for two to three years and sometimes longer although at a diminished laying rate.

A mated queen can produce both fertilised and unfertilised eggs. Fertilised eggs produce workers or may be used to produce new queens, while unfertilised eggs produce drones. An unmated queen will only produce drones. How the queen differentiates

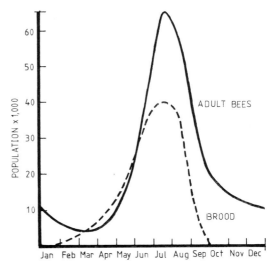

Fig 2 The breeding cycle of adult bees and brood, showing the July peak

between laying a fertilised egg in a worker cell and an unfertilised egg in a drone cell is still unexplained, although there are a number of theories. One thing is certain, her laying rate is quite phenomenal and may reach two thousand eggs a day at the peak of the summer.

The worker bee

Worker bees are underdeveloped females or neuters, and it is normal to refer to them using the female gender. The worker bees, as their name suggests, do all the work. They secrete wax and build comb with it, gather stores of pollen and honey, rear the brood, ventilate and protect the hive. The worker bee is armed with a sting which it will use in the defence of the hive. This sting is barbed and injects a poison which is a compound venom, unlike the simple alkaline irritant of the common wasp. The worker bees regulate the queen's egg-laying activities, determine when new queens will be reared, build drone cells to enable drones to be bred and evict those same drones in the autumn. It is the worker bee that is the real power in the hive.

A newly hatched worker bee has a downy, soft appearance, but this soon wears off and to human eyes she becomes indistinguishable from her older sisters in the hive. There are clearly defined duties for a worker dependent

17

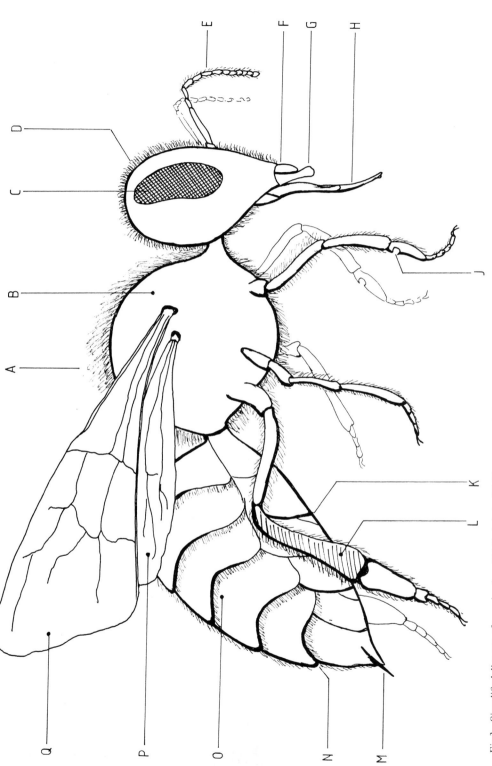

Fig 3 Simplified diagram of a worker bee: (A) Body hair. (B) Thorax; containing muscles for locomotion. (C) Compound eye; main source of vision. Bees appear to have the ability to recognise some shapes and colours. (D) Simple eyes; three located on top of the head. (E) Antennae; sensory devices that transmit environmental details to the brain, particularly scent and taste. (F) Labrum (upper lip). (G) Mandible (jaw); particularly useful for moulding wax. (H) Proboscis (tongue); used to suck up liquids. (J) Antenna cleaner. (K) Wax glands; located under the bee's abdomen, and used to secrete wax for building comb. (L) Corbicula (pollen basket); used for transporting pollen. Considerable amounts of pollen can be carried in this. (M) Sting; although drawn partly exposed, the sting is retractable. (N) Nasonor (scent gland); exposed by turning down the last segment of the abdomen. (P) Hind wing. (Q) Front wing

upon her age. First duties consist of cleaning the cells and then acting as a 'nurse' bee to the larvae, feeding them and keeping the brood warm. Later duties include feeding and grooming the queen and building comb or queen cells when required.

The first fourteen days or so, depending on the time of year, are spent on this 'hive duty'. This is not to say that a worker does not leave the hive during this time. Cleansing flights are common and are especially noticeable in the spring when a large number of bees will leave the hives on a sunny morning to empty their bowels. After about two weeks the worker bee concentrates on collecting nectar and pollen. This is the so-called field bee or flying bee stage of her life, and is the last. As well as collecting nectar and pollen, the flying bees gather propolis, a resinous tree gum.

The life expectancy of the worker bee depends on the season. Bees hatched in the late autumn may survive through to the spring, giving a possible life span of some five months, but in mid-summer the life span is very short. A worker bee who survives all accidents and enemies can only be expected to live for five or six weeks. They literally wear themselves out because they operate on instinct, not reason. The older bees in the hive can usually be distinguished by the condition of their wings which become frayed and ragged; and the motive power to the wings is possibly weakened after three or four weeks of lugging full loads of nectar and pollen back to the hive. A worker bee will leave the hive one day to load up with nectar and pollen and never return. Perhaps if a bee could reason it would carry reduced loads and make more journeys, thereby surviving for a few more hours or even days, but this is not the way of the bee. When viewed in this context, that often-quoted saying 'as busy as a bee' takes on a new and sombre meaning.

The drone

The drones are the male bees and are formed from unfertilised eggs. Drones do not possess a sting and have no specialised function except to mate with available virgin queens. They are therefore indispensable, although drones are non-producers and do little to support themselves and nothing to assist the daily functioning of the colony.

Drones are often referred to as being fat and lazy; this is a mistake because human failings cannot be attributed to bees. They are much larger than the workers, being bred

A shallow super frame. Here the queen found her way past the excluder, and laid eggs in the middle of the frame. After placing the queen back in the brood chamber, the drones from these cells were removed and destroyed *(author)*

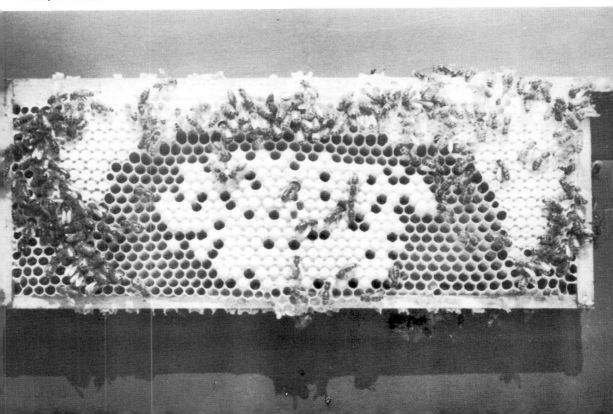

A Hoffman frame containing an expanding area of brood, well covered with bees *(author)*

A Hoffman brood frame from the outer edge of the brood nest, fairly well fitted with stored honey *(author)*

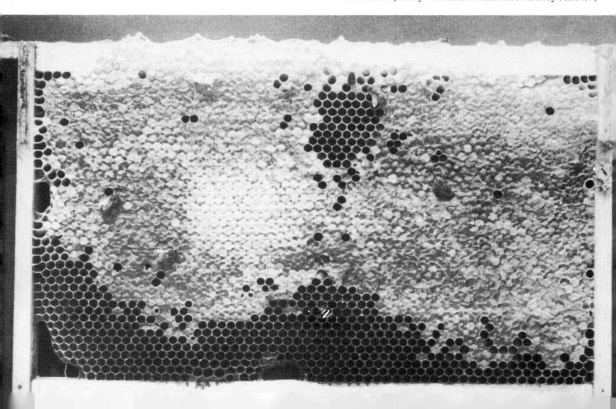

in bigger cells, but are shorter and stockier than the queen. They present a clumsy and, I think, almost comic appearance with their large compound eyes and distinctive low-pitched buzzing when in flight.

The drone is a very powerful bee in flight and this is probably necessary as the act of copulation with the queen takes place in mid-air. The virgin queen is pursued by a considerable number of drones and sometimes other bees which seem to get caught up in the excitement of the action. A drone in the act of mating loses his sexual organs and consequently his life.

Drones are normally present in a healthy colony of bees from late April to August. Those drones that have survived through the summer are forced out in the autumn by the worker bees. Sometimes they can be seen clinging around the hive entrance at evening time; if the cold does not kill them, then they starve to death. Drones can be present in a colony until later in the year if there are unhatched queen cells for any reason; but, like the queen, their actions and life span are determined mainly by the worker bees.

Bee stings

Bees possess a sting as a deterrent to intruders and robbers. It is rarely used except in defence of the colony, and when used against human flesh results in the death of the bee. Unlike the simpler pointed sting of the wasp, the sting is barbed and cannot be withdrawn by the bee from human skin which is elastic in nature. As the bee struggles to free itself, the sting, usually complete with poison sac, is torn from the bee's body. Often the doomed insect will continue to try and sting although it no longer has the means to do so.

We are all responsive to bee stings, and anyone who handles bees will eventually get stung. For most beekeepers, stings are merely a nuisance to be tolerated and avoided whenever possible, although there are a large compound eyes and distinctive low-a mild allergy to the sting.

If, or rather when, you are stung there is little that can be done to relieve the pain, although there are bee-sting antidotes on the market usually in the form of externally applied liquid or paste. I have only slight experience of such antidotes and do not feel qualified to comment on their effectiveness, but one consolation is that, in general, the more one gets stung the less the effect. The venom is absorbed by the system and eventually provides a partial immunity. Some beekeepers I know claim that stings are little more painful or longlasting to them than a nettle rash. When stung, it is beneficial to remove the barb and poison sac as quickly as possible as it will continue to pump poison into the wound for several seconds. Also be aware that when a bee stings or is accidentally crushed it excites other bees to attack. This accumulative response can prove very interesting when dealing with an aggressive colony.

Some bee behaviour

Bees operate instinctively and do not, as far as we know, possess the facility to reason. Therefore, except for slight differences in 'character' between individual colonies, ie mild or aggressive nature, inclination to swarm or otherwise, bee behaviour in response to situations or stimuli is the same. The newcomer to beekeeping will probably not observe, or at least not recognise, a 'bee dance' but fanning, swarming and the activities of guard bees are easily observed.

Fanning is used in two situations: either to draw air through the hive for cooling or to attract other members of the colony. Strictly speaking, these are distinctly different operations. When the weather is warm a large number of bees will operate their wings in a certain manner whilst stationary in order to draw air through the hive. This assists cooling and is also thought to aid evaporation of the stored nectar. Excess water evaporating from the stored nectar would soon create a very humid atmosphere if there were not a flow of air through the hive.

What is normally referred to as fanning is that activity in which a bee assumes a head-down, tail-up position and again operates its wings whilst remaining stationary. If one looks carefully at a bee in this position, there seems to be something amiss with its abdomen. What the bee does is turn down the last segment of its abdomen to expose the scent gland. Fanning its wings spreads the scent to the other members of the colony. Such fanning is particularly noticeable when hiving a swarm; the first bees to find the entrance of the hive will form ranks around the entrance and start fanning. The rest of the bees will follow the source of this scent, thereby arriving at the hive entrance. To induce fanning artificially, move the hive slightly or, even better, change the position of the entrance. Bees returning to the hive will be temporarily confused and initially fail to find the entrance. Those bees that find it first will usually start fanning to attract other returning bees. In human terms it is a case of 'Hey chaps, I've found it. Come over here.'

A beekeeper uses the term 'guard bee' to describe those bees which defend the entrance of the hive when danger threatens. This is not a regular function carried out by certain bees but an instinctive communal response to interlopers. Should there be some disturbance, a number of bees will move to the entrance ready to challenge or repel intruders. Intruders may be other bees or wasps intent on robbing or, of course, the beekeeper.

If you wish to test the efficiency of a colony's defence system, then approach the

Bees in typical fanning posture at the entrance to the hive *(author)*

hive on a warm quiet summer evening. If everything is in order then there will be no guards at the entrance, just foraging or field bees drifting in and a few younger bees exercising. Tap the hive and one or two bees will appear at the entrance. A few more taps and a fair number of bees will appear. Some may run forward if there is a flight board fitted and one or two take to the wing and circle around. In the autumn when wasps and other bees are intent on robbing a hive of its contents, the defence of a hive is of vital importance.

There are occasions, however, when guard bees will let a strange bee enter the hive. If a bee strays to another hive, perhaps because the hives have been moved and it has mistakenly chosen the wrong one, then it will be allowed to enter, especially if it is loaded with nectar and pollen. Now, a bee does not recognise a stranger to the colony by sight but by smell. Each colony has its own odour which is not discernible to humans. A strange bee which believes it is entering the correct hive will be challenged by the guard bees and perhaps 'roughed up' a little, but as the incoming bee does not exhibit the aggression of a robber it will not be subject to a 'life or death' attack. By the time a few guard bees have examined the incoming bee, it will have begun to take on some of their odour, the

Brace comb and natural comb built on the bottom bars of super frames, due to incorrect spacing *(author)*

colony scent. From then on the bee is accepted into the colony and can come and go at will. If, however, it should on leaving that hive find its original home a few feet away then the whole process of challenge and acceptance will be repeated because of the different odour that the bee will now be carrying.

Both fanning and guarding are functions easily observed, but bee dances are little understood except the bee food dances. Consider the problems of communication without speech. We humans have developed a sign language and bees have a similar language, perhaps several thousand years old. If a bee chances upon flowers rich with nectar and pollen, it returns to the hive loaded with these goods. It is not by chance that other bees from the hive find the same flowers. The foraging bee will perform one of two types of dance commonly known as the 'round dance' and the 'wagtail dance'.

The round dance is used when food is to be located within the immediate vicinity, say within a 90–100yd (80–90m) radius of the hive. The wagtail dance is used for a food supply exceeding this distance. While the round dance means 'there's food out there, go and get it', the wagtail dance is much more specific. The rate at which the bee dances, the length of time spent on the wagtail section of the dance and a number of other factors spell out to the other bees a directional bearing relating to the angle of the sun. As far as we know, it does not determine the type of blossom, but during each dance the dancing bee will pause and hand out samples of pollen or nectar so those bees leaving the hive know what food to look for.

This dancing is remarkably complex, and scientific experiment has shown it to be very effective. What is quite amazing is that these dances are usually held in the dark on the face of vertical hanging comb but still the message gets through.

Propolis

Propolis is a dark-coloured gummy resin gathered by the bees mainly from trees that excrete resin or have buds with a resinous coating such as horse chestnut. The bees use propolis to fill up cracks and most other gaps

23

Natural comb built under the crownboard of a hive. It is interesting to note that under certain conditions bees will build natural drone foundation for storing honey. The comb shown was almost exclusively used for this purpose *(author)*

that are smaller than a 'bee space'. I have examined the remains of a once long-established colony in a hollow tree and the insides of the cavity were smooth and dark where propolis had been liberally applied.

When the weather is hot, propolis is soft and sticky and something of a messy nuisance when examining the hive. A beekeeper's gloves or gauntlets soon become stained with propolis, and smokers and hive tools all carry evidence of contact with it. As the weather cools so the propolis becomes harder, eventually achieving a glass-like fragility; then it will shatter and fly in all directions if any attempt is made to move it.

Some colonies of bees are more adept at collecting propolis than others. Hives standing side by side in an apiary may contain widely varying amounts. The time of year, location and weather conditions will also affect the amount of propolis collected. I have owned bees that only apply propolis sparingly, while others have used it so extensively that it dribbles down the inside walls of the hive.

Brace comb

In any hive, gaps which are larger than 'bee space' will be filled by the bees with extra comb. For instance, if the brood chamber is too large and the gap between the brood chamber wall and the edge of the hanging frame exceeds $\frac{3}{8}$in (10mm), the bees will use this space to build extra comb. It is often not constructed to a regular pattern and is known as brace comb. It is not the same as a full natural comb which the bees will build, hanging from the crownboard, if you leave a frame out.

Brace comb will be built at any angle from the main combs and even along the top of the hanging frames if the gap under the crownboard exceeds 'bee space'. It really does have the appearance of bracing the whole assembly, and brace comb and propolis together can turn a hive full of frames into a remarkably rigid structure.

3 Hives

The bee hive used in most countries today is basically a wooden box in which rectangular wooden frames carrying the wax comb hang. Before the introduction of this movable (or more correctly removable) frame hive, primitive constructions containing natural comb were used. The most popular device was the skep, which was shaped like an open-ended dome and constructed from straw bound with split bramble.

The skep has survived in Britain, although in very limited numbers. The decline of skep beekeeping in this country began around the turn of the nineteenth century, and those examples currently in use are often kept only for educational purposes or out of curiosity. There are a few individuals promoting the revival of the skep, but although it is an

adequate home for bees, it is not a suitable device for exercising control over them. Examination of the contents is difficult and a thorough examination with the bees *in situ* is impossible. A number of beekeepers possess a skep and use it for collecting swarms; this is probably as good a use for it as any. However, unless you have a particular desire to own a skep, a simple wooden box will suffice for swarm collection.

The movable frame hive

All hives are simply man-made artificial homes for bees. Bees are hived for convenience and the need to control or confine their activities. It is little different in principle from enclosing cattle in a field. The theory and eventual adoption of the movable frame hive is, however, almost as far

Traditional skeps used for keeping bees, rarely used in Britain today *(by kind permission of Mr Karl Showler)*

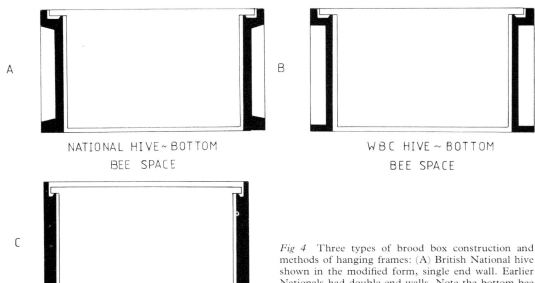

NATIONAL HIVE ~ BOTTOM
BEE SPACE

WBC HIVE ~ BOTTOM
BEE SPACE

SINGLE WALLED BOX HIVE
TOP BEE SPACE

Fig 4 Three types of brood box construction and methods of hanging frames: (A) British National hive shown in the modified form, single end wall. Earlier Nationals had double end walls. Note the bottom bee space and long lug top bar. (B) WBC hive showing typical lightweight construction, bottom bee space and long lug top bar. (C) Typical single-walled box hive showing robust construction, top bee space and short lug top bar

removed from the methods of skep management as the modern farmer is from our hunting ancestors.

In the natural state bees construct wax comb that hangs vertically. The main anchorage is along the top, although in a confined space, such as a small hollow tree, it may also be supported along the edges by brace comb. The movable frame hive duplicates the hanging comb but each piece of comb is fixed in a rectangular wooden frame. The frames, usually ten or eleven in number, are not fixed but hang by projecting lugs in an open-ended box or brood chamber. The brood chamber is closed at the top by a crownboard, which is simply a loose wooden lid, and at the bottom by the hive floor. Entry and exit of the bees to and from the interior of the hive is controlled through an entrance block or adjustable shutters.

Surplus honey, ie the honey removed by the beekeeper, is stored in shallow frame boxes known as supers or in section racks. Although it was standard practice at one time to use a super and section rack, it is now much more common to use multiple supers. The whole assembly is topped by a removable wooden roof covered with zinc sheet which forms a durable waterproof shelter against the elements.

Bee space

The fundamental discovery which made the movable frame hive a practical proposition was 'bee space'. This is a dimension, ideally of $\frac{1}{4}$in (6mm), which bees will use as a passageway. In practice, a gap of $\frac{1}{4}-\frac{5}{16}$in (6–8mm) will suffice. Bees will not block a bee space gap with brace comb or propolis. Under the crownboard, for example, there is a bee space gap which allows the bees to travel across the top of the hanging frames. If that gap were to exceed $\frac{5}{16}$in (8mm), the bees would extend brace comb from the top of the frames to the crownboard. If the gap were less than $\frac{1}{4}$in (6mm), then the bees would fill it with propolis and glue the whole assembly together making an awful mess. Consequently, the internal fittings of a hive are constructed either to make direct contact or to maintain bee space.

Types of hive

Both single-walled (box) hives and double-walled (WBC-type) hives are presently in use in Britain. The double-walled hive, although the older design, is in effect a box hive with a separate outer casing of detachable 'lifts'. Many amateurs use the double-walled hive, possibly because they are abundantly available secondhand. They were for many years regarded as the ideal British hive, but an increasing number of amateur beekeepers now make use of the simpler, less traditional, box hive.

The epitome of the double-walled hive,

Cutaway of the modified British National hive, showing the brood chamber, shallow super box and section rack (*E. H. Thorne (Beehives) Ltd*)

the WBC pattern, was invented at about the turn of the century by W. B. Carr. The success of this design was such that most double-walled hives are now referred to as WBCs even though some may not be true to the original design. The traditional WBC hive was undoubtedly designed with aesthetic appeal in mind, and it does look very attractive with widely overhanging flight board and gabled roof.

There have been other double-walled hives which are no longer marketed, but may still be encountered. These were 'Cottager' hives, a sort of poor man's WBC, 'Combination' hives, 'Conquerors' and 'Double Conquerors', plus many others of differing

Table 1 Details of hive types

Hive	Type	Maximum number of brood frames	Brood frame size	Super frame size
British National	Box, long lug frames, bottom bee space	11	$14 \times 8\frac{1}{2}$in (35×21cm)	$14 \times 5\frac{1}{2}$in (35×14cm)
Smith	Box, short lug frames, top bee space	11	$14 \times 8\frac{1}{2}$in (35×21cm)	$14 \times 5\frac{1}{2}$in (35×14cm)
WBC	Double-walled, long lug frames bottom bee space	10	$14 \times 8\frac{1}{2}$in (35×21cm)	$14 \times 5\frac{1}{2}$in (35×14cm)
Modified Commercial	Box, short lug frames, bottom bee space	12	16×10in (40×25cm)	16×6in (40×15cm)
Langstroth	Box, short lug frames, top bee space	10	$17\frac{5}{8} \times 9\frac{1}{2}$in ($44 \times 24$cm)	$17\frac{5}{8} \times 5\frac{3}{8}$in ($44 \times 13.5$cm)
Modified Dadant	Box, short lug frames, top bee space	11	$17\frac{5}{8} \times 11\frac{1}{4}$in ($44 \times 28$cm)	$17\frac{5}{8} \times 6\frac{1}{4}$in ($44 \times 16$cm)

styles. Some 'hives' were even made to house skeps, although such transitional designs were not produced for many years.

The Americans were responsible for the introduction of the simply constructed single-walled box hive. This type of hive is not usually regarded as visually pleasing and does not have the 'garden ornament' attraction of the double-walled WBC-type. Single-walled hives are just boxes and there are a number on the market of different dimensions. The simplest type has rebates machined in two opposite sides to accommodate the top bar lugs of the hanging frames. Most box hives use frames with 'short lug' top bars, although the British National hive uses frames with 'long lug' top bars which require a slightly more complicated hive construction.

The Langstroth and Modified Dadant are both box hives of American origin; the Langstroth is claimed to be the world's best-selling hive. In Britain, the British National hive, commonly referred to as the 'National', is used extensively, while the Smith hive, which requires the same frames with shortened top lugs, is something of a favourite with the home constructor. The Modified Commercial is, as its name implies, intended largely for the commercial honey producer; a useful feature is that some of its component parts are interchangeable with the National hive.

Apart from the use of short or long lug top bars, the other design feature to note is whether a hive is constructed with top bee space or bottom bee space. In top bee space hives, the frames, when fitted, lie approximately $\frac{1}{4}$in (6mm) below the top edges of both deep (brood) and shallow (super) boxes. In bottom bee space hives, the frames fit flush with the top edges of the brood and super boxes, bee space being provided beneath the frames. Top bee space hives use crownboards with a plain flat side over the brood chamber or super, and are usually reinforced along the top (upper) edges. Bottom bee space hives use crownboards with $\frac{1}{4}$in (6mm) bee space over the brood chamber or supers, provided by framing it with strips of wood $\frac{7}{8} \times \frac{1}{4}$in ($21 \times 6$mm) thick. Table 1 lists the types of hive available, together with their characteristic features and size and number of frames.

Nucleus hives

It may not always be necessary or desirable to use full-sized hives. Nucleus hives are designed to take from two to five frames and are primarily intended for breeding purposes. They are usually used in conjunction with

Modern glass-sided observation hive *(by kind permission of the British Beekeepers Association)*

full-sized hives, so it is essential to employ a nucleus hive which accepts the same size frames as other hives in the apiary. Two-frame nucleus hives are used for queen rearing, although I do not think any such devices are currently listed by manufacturers of beekeeping equipment. Most nucleus hives are designed to hold five frames which approximates to half a full colony, and these may be purchased ready-made or in the flat, although one may be limited to only a few frame sizes.

Most nucleus hives are like all single-walled hives, simple boxes; although some pretty little hives modelled on the WBC-style can be found albeit without the double-wall feature. Many nucleus hives are made by the home constructor of comparatively flimsy materials and with fixed floors. This does not matter as they are normally only used in the summer when the weather is kindest and the hive is occupied for a short term so that fouling of the floor is not a problem. Unfortunately, most homemade nucleus hives have insufficient space under

the roof for a feeder of a decent size; a point worth looking at when purchasing one or to consider when making one's own.

Observation hives

Nowadays, the observation hive is a glass-sided frame containing a brood frame and two supers. These are used for educational purposes and may be found in schools or at agricultural shows where there is a beekeeping exhibition. Most of these hives swivel so that both sides can be examined at leisure. If used inside a building, a polythene tube of about 1¼in (3cm) diameter is used from the entrance of the hive to a point outside the building to enable the bees to come and go without causing a nuisance.

The principle of observation hives is not new and they have not always followed the pattern described. Since the introduction of the movable frame hive, there have been individuals keen to study the lifestyle of the honey bee and the internal workings of the hive. Many of the earlier observation hives were conventional-type hives with obser-vation panels. Some were quite ingenious, beautifully constructed and bear little

29

resemblance to the device we recognise nowadays as an observation hive.

Hive materials and weatherproofing

The British Standards Institution recommends Western Red Cedar for the construction of wooden hives. Apparently, this wood does not require any protection but, bearing in mind the present value of a hive, an application of a suitable preservative would not come amiss. With double-walled hives, only the outside lifts, roof and floor need to be constructed of a durable timber as the brood chamber and supers are not exposed to weathering.

Although a hive constructed from Western Red Cedar may last some fifty or sixty years, other materials can be and have been used. Many soft woods are unsuitable because they will shrink and warp to an unacceptable degree, and it is difficult to obtain timber suitable for the full depth of a deep brood box. One alternative is to use an exterior

grade plywood. I have used a National-type hive made from $\frac{7}{8}$in (21mm) plywood obtained from a building site. It required a good coat of paint for protection but otherwise served as well in the short term as the proprietary hives in the apiary.

A combination of plastic and wood has been used for hives as has polyurethane foam. This material has the advantage that it is lightweight (although this can be a disadvantage on a windy day) and is resistant to weathering and rot. This type of hive is manufactured so that all parts interchange with the more conventional wooden construction. Until recently, the difference in price between the two types was minimal and insufficient inducement to change from the proven wooden hive.

It is normal practice to cover the roof of a hive with zinc sheet, and this is an excellent weatherproof and durable arrangement. If such sheeting is not available, roofing felt is a good substitute which will last several years. A double-walled hive with a gabled roof can be fitted with two asbestos roofing 'slates' with a wooden strip at the apex to cover the join. This has the advantage of adding weight to the roof making it less likely to be blown off in high winds.

There are three preservatives normally used for wooden hives: paint, creosote and

One way of buying beekeeping equipment. A sale of secondhand equipment, organised by a local branch of the British Beekeepers Association. There are traditional white-painted WBC hives in the foreground, with nucleus hives next to them *(by kind permission of Gloucestershire College of Agriculture)*

Cuprinol. Traditionally, double-walled hives were painted white or cream but this is not essential. It is often advisable to paint hives, if they must be painted, a colour which blends in with the surroundings, particularly when hives are situated some distance from habitation (known as out apiary) where vandalism can be a problem. A white or cream hive is easily visible from a quarter of a mile away, while a hive painted green standing against a hedgerow is almost invisible at the same distance. Creosote is a useful preservative but a hive so treated must be weathered or allowed to dry thoroughly until the strong smell disappears. A 50:50 mixture of creosote and paraffin dries quickly and the smell does not linger so long, but this is not the best weatherproofing. The preservative marketed under the brand name Cuprinol, however, is excellent. It can be purchased colourless or with a stain incorporated, and is manufactured by Cuprinol Limited in Frome, Somerset. It does not matter which you use; the weather-resisting properties are the same. The great advantage of Cuprinol is that it dries quickly, without smell, and hives treated with it can be used within a few hours of treatment.

Temporary hives

During the summer months the need often arises for an extra hive or hives. This may be to hive swarms, conduct queen-rearing experiments or to make up a few nuclei for selling. Whatever the reason, it is not necessary to use hives made of the best quality timber as they will not normally be exposed to the more inclement weather of autumn and winter. As long as temporary hives conform to the required frame size and are sufficiently beeproof, the bees will do as well as those kept in more conventional hives.

Plywood, cardboard, chipboard and possibly polystyrene are all suitable materials for temporary hives. However, some materials should be avoided, especially those which are subject to condensation such as plastics and polythene. In most cases, temporary hives are only needed for part stocks of perhaps four or five frames and then,

because of the reduced dimensions, they will be fairly rigid even when constructed of comparatively flimsy material.

Temporary hives often need to be constructed quickly to cope with an unexpected situation. It is far better to use a temporary hive, perhaps of cardboard, than to resort to hiving swarms in boxes or skeps. I have used cardboard for nucleus hives based on the required dimensions to accept five British Standard frames. The cardboard used was not of the 'shoe box' variety, but industrial $\frac{3}{8}$in (10mm) double-thickness corrugated packing-case cardboard. It is cheap (usually free if secondhand), light, remarkably strong and seems to have good insulating properties. These hives were held together with 2in (5cm) brown parcel tape and two thin strips of wood fixed to the floor to stand them clear of damp. One such hive stood outside for most of the summer of 1980 without any external protection and served very well, although the floor had become rather damp and tatty by the autumn.

Plywood for a hive need only be $\frac{1}{8}$in (3mm) thick, although this will require some reinforcing at the corners and, for maximum strength, the floor should be fixed. A fixed floor is essential for most temporary hives to give sufficient rigidity to what is, in essence, an open-ended box. I have not found it necessary to construct nucleus hives of anything other than thin plywood. As long as they are well painted, the weather does not penetrate.

Chipboard is useful but has the disadvantages of being heavy and of expanding alarmingly if subject to excessive damp which affects the internal bee space. The weight should also be considered as a hive of bees is heavy enough to transport without adding unnecessarily to it. Where chipboard is particularly useful is in the form of a floor. A piece of chipboard $\frac{1}{2}$in (12mm) thick as a floor adds considerably to the strength of a hive constructed from $\frac{1}{8}$in (3mm) plywood.

I have not yet tried polystyrene sheet, but I doubt whether it would be suitable for the complete hive. It could be useful, however, in conjunction with cardboard or thin plywood, and a sandwich of polystyrene and plywood would make a strong and well-insulated hive which would probably be

31

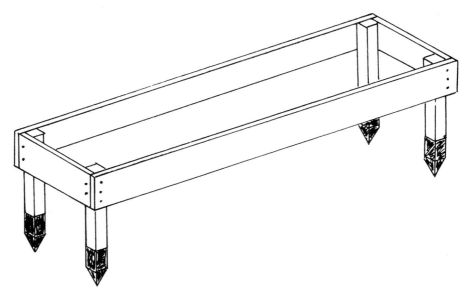

Fig 5 Simple hive stand for two hives. The stand should be about 4ft (120cm) long × 14in (35cm) wide, and no more than 18–20in (45–50cm) high. As shown, the side and end rails are made of 6 × 1in (15 × 2.5cm) timber, which gives sufficient rigidity (on a stand this size) without the need for further supports

suitable for winter use if it were given some protection from the worst of the weather.

A temporary hive can also be made by using the internal boxes of a double-walled hive of the WBC-type. A simple floor after the pattern of a box hive can be made and a roof of similar type. If standing inside a shelter, the roof can be dispensed with, the crownboard being sufficient protection. Using a number of manufactured parts has the advantage of reducing the time needed to make a temporary hive, which is often required in a hurry; availability is therefore an important factor. At present, brood boxes and supers for double-walled hives are cheap and plentiful on the secondhand market. This is because the outside lifts on a double-walled hive suffer all the weathering and consequent decay, leaving the internal structure untouched except perhaps for attack by woodworm. There have also been a vast number of WBC-type hives produced over the past sixty years in this country whereas box hives have only been accepted by the amateur fraternity since World War II.

Accommodation for bees, therefore, need not be expensive; and there are probably other materials or combinations of materials which can be used in addition to those I have described. If you experiment you will at least cause comment. The cardboard hives I used were frowned upon by some, but as long as you are satisfied that the materials used will cause no harm to the bees and they are kept warm and well fed that is all that matters. As long as these conditions are met, they will do no better in the most expensive hive that could be constructed.

Hive stands

WBC hives with fitted legs do not need a stand, although it is useful to support them clear of the ground if possible. This has the dual advantages of bringing them to a better working height and helps prevent the legs and lower hive floor from rotting as a result of moisture drawn up from the ground. This type of hive always looks neat placed on a concrete paving slab sunk into the lawn, but during the wetter times of the year the legs and bottom of the hive can become soaked with rainwater splashing up from the slab.

Box hives do not have an integral stand and so it is necessary either to build a stand or find a substitute such as four housebricks or an old milk bottle crate. A plastic milk bottle crate makes a particularly useful hive stand, but obtain them from your local dairy, who often have a few broken crates of no use for carrying milk bottles, but quite adequate for a beekeeper's purpose. Do remember not to help yourself to milk bottle crates—that is stealing! A fixed wooden construction is the best in a garden site or permanent apiary. It should be built so that the top run of timber is about 18in (45cm) off the ground. This puts the hive at a good working height for most people. The legs should be well soaked in creosote to a line above ground level for it is at that point that they will rot. The remainder of the stand can then be treated with a preservative in the same manner as the hives.

4 Appliances and Accessories

Beekeeping as a hobby is little different from gardening, painting or home car maintenance inasmuch as there is an active market retailing appliances and accessories. Most of these items are either a necessity or a real aid to handling bees, but there are also non-essential items which range from the useful to the frivolous. Mercifully, beekeeping is fairly free from the latter.

As it will be useful to describe them in greater detail I have included crownboards, frames and wax foundation in this chapter, although they are not accessories but an integral part of the movable frame hive.

Frames

Frames are rectangular devices, designed to hold the wax comb. Whatever the type of frame, they all hang by the top bar which has projecting lugs. Most hives now use the short lug, which simplifies hive design as already outlined in Chapter 3. The long lug is peculiarly British and is really a throwback to the early movable frame hives which used separate metal end spacers.

There are two basic methods of spacing frames. The oldest employs separable metal ends, which are made of folded tinned steel, although these are gradually being superseded by plastic of similar design. A far superior method of frame spacing is the self-spacing Hoffman frame. The side bars of the Hoffman frame have winged extensions at the top which butt against each other to give the required spacing. One edge of the winged extension is flat while the other is chamfered to a point or rounded. This makes it easier to part when the bees have tried to cement the frames together with propolis. The chamfered or V-edge of the Hoffman frame should always be on the left when the frame is viewed the right way up from either side.

There are other methods of frame spacing which have their uses, including plastic devices currently marketed under the name of 'Yorkshire spacers'. These convert parallel sidebars to the Hoffman-type of self-spacing. They are preferable to metal or plastic end spacers and have the same advantages as the Hoffman frame over the end spacers: frame spacing is more precise and newly inserted frames, which have not been 'propolised' by the bees, do not shuffle about or swing when the hive is moved as they do with end spacers. In addition, if metal or plastic end spacers are used in a super they have to be removed for extracting purposes while the Hoffman or Yorkshire-spaced frames are used as they are.

Now, while metal or plastic end spacers, Hoffman frames or Yorkshire spacers can be used in brood chambers or supers, it is probably extravagant to use them in the supers as they are more expensive to purchase then parallel-sided frames. A simple and effective method of frame spacing can be achieved by using slotted or castellated spacers. This type of spacer fits across either end of the brood chamber or super and both supports and spaces the

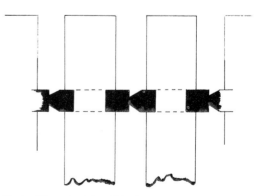

Fig 6 The principle of Hoffman spacing. The frames shown are the British standard long lug type used in WBC and BS National type hives

frames. It is a method which I am particularly fond of for use in National pattern supers, and it has the advantage of precise spacing while holding the frames fairly rigid. At present, it is slightly more expensive to use castellated spacers and parallel-sided frames than Hoffman self-spacing frames. In the long term, however, the spacers will last the life of the hive, becoming an integral part of it, and will still be in use when there have been several changes of frames.

For one season I used castellated spacers in the brood chamber of a hive which convinced me not to do it again. It is not often that one needs to swap or move frames in a super but frequent manipulation of the frames is required in the brood chamber. With castellated frame spacers, each frame has to be attended to individually. It is not possible to remove an end frame and slide the others apart to facilitate removal or break the grip of the propolis. I may sound finicky, but once you have tried manipulating the frames in the brood chamber of a hive you will appreciate my point.

Another method of frame spacing worth

Empty British Standard super and brood frame, showing the comparison of sizes. The wax in the large brood frame has become considerably blackened with use. The brood frame is the Hoffman self-spacing type, whilst the super frame is not fitted with spacers because it is used with castellated spacers fitted in the super box (*Jak Photography*)

consideration and confined exclusively to use in supers is the Manley-type self-spacing frame. These frames are close-ended; in other words, the sides, when assembled in the hive, make contact along their full length. These frames do not require a V-edge as they do not have to be moved as frequently as those in the brood chamber and, as supers are only on the hive for part of the year, they never seem to suffer such an extensive application of propolis as the brood chambers.

For the absolute beginner, the most confusing section in a catalogue of beekeeping equipment is probably that part dealing with frames. It soon becomes obvious that there are deep and shallow frames to suit brood chambers and supers, and then there is a variety of quoted dimensions and a choice of wedge top or sawcut top bar frames. It is, of course, simpler than it first appears once you have some idea of your requirements. Most British catalogues list frames in two sections. The first deals with those frames made in accordance with British Standard specification BS1300, 1960, both deep (prefix DN) and shallow (prefix SN) frames which fit most of our double-walled hives and the British National hive. The second section lists frames suitable for other hives such as the Modified Commercial, the Modified Dadant, the Langstroth and a few variations on these. Most of these frames are self-spacing, ie Hoffman-type, with what is called the 'wedge' top bar.

The wedge top bar is the easier type of frame to construct, especially if ready-wired wax foundation is being used. It is a little more difficult to fix the wax foundation in the sawcut top bar frame and it is gradually being superseded by the wedge-type. Some manufacturers of beekeeping equipment no longer list the sawcut top bar, and with the self-spacing Hoffman frame only the wedge-type top bar is available.

It is normal, although not essential, practice to space the frames in a super wider than those in the brood chamber. With end-spaced frames the solution is simple; there are narrow spacers of $1\frac{3}{8}$in (3.5cm) for the brood chamber and wide spacers of $1\frac{7}{8}$in (4.6cm) available for the supers. Similarly, the slotted spacers are arranged for narrow and wide spacing. If you decide to use Hoffman spacing throughout, you may well find that only the narrow spacing is available for both deep and shallow frames. Still, there is no reason why you should not 'mix and match'. Use Hoffman self-spacing frames for the brood chamber and some other arrangement for the supers. It is, of course, a matter of personal choice, but I think there is little doubt that the best arrangement for brood chambers is the Hoffman frame and it is well worth the small extra cost.

An alternative to the wooden frame is the plastic frame, of which there is a limited number of types available at present, including the British Standard shallow which requires end spacing and the British Standard deep and shallow with Hoffman-type self-spacing. I purchased a few of these for evaluation but did not put them to use in the apiary. The sample frames that I examined were bowed to the extent that when the frames were dropped into a hive the centre of the frame was some $\frac{1}{8}$in (3mm) proud. This reduced the bee space between the frames and crownboard to the point where the bees would fill the gap with propolis. Consequently, they were never used and I cannot comment on their practical usefulness, but they were certainly robust and it is simple to fit in the wax foundation. The frames part along the centre line, the foundation is laid on one half and the other half located to it on fixed pins. They hold together very firmly and eliminate the need for pins and glue. Another advantage, so the makers claim, is that unwired foundation can be used as there are built-in reinforcing strips equally spaced across the frame.

I feel sure that such items as plastic frames will, in the future, be standard fittings and the wooden frame will be a rarity. However, until the plastic frame is either better made or very competitively priced, I cannot see the beekeeping fraternity flocking to purchase them in large numbers.

Wax foundation

The beekeeper does not normally introduce natural wax comb into the hive frames but uses thin sheets of beeswax in which the base

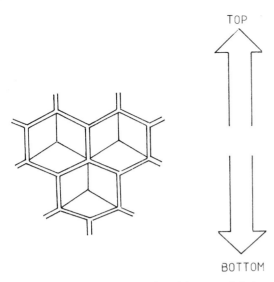

TOP

BOTTOM

Fig 7 There is a wrong and a right way of fitting foundation to a frame. This figure shows the way foundation should be hung

of each cell is already formed. These wax sheets, which may be pressed or moulded with either worker- or drone-sized cells, are drawn out to full cell depth by the bees. It is then virtually indistinguishable from natural 'wild' comb. The bees have, in effect, been given a foundation to work from.

It is normal practice to use a size of wax foundation the same size as the frame being used. At one time starter strips were used: thin strips of foundation fixed across the top of the frames. The bees would start to draw these out and then expand downwards in the normal manner until the bottom bar was reached. I have tried this method when I was short of foundation, cutting full frame sized sheets longitudinally into three. Each third of a sheet was fixed into the top of the frame in the normal manner and the bees left to draw it out. It took them perhaps a little longer to complete than a full sheet of foundation, but the finished product was quite neat even though the lower corners were not completed.

There are various sizes of wax foundation available cut to suit most, if not all, sizes of brood and super frames. A specially thin foundation is available for producing sections and cut comb, both methods of honey production that avoid the complication of extraction and bottling. Deep (brood) and

shallow (super) foundation may also be obtained with wire reinforcing which greatly strengthens a comb of brood. It is unwise to use brood frames without wiring. The wiring stops the wax sagging and an unwired comb which can contain several pounds of brood, honey and pollen may fall apart if mishandled. The same is true of shallow super foundation when extracting honey. A centrifugal extractor, especially the tangential type, places considerable strain on the comb. Wire bracing can make all the difference between a cleanly extracted comb or having one fly to pieces. It is possible and was, at one time, almost universal practice, to wire sheets of foundation. On a commercial scale I understand savings may still be made, but I would advise anyone operating on a small scale to pay a little extra and purchase foundation ready wired.

An alternative to natural beeswax foundation is plastic foundation which is produced in the USA, Australia and Great Britain. Of the brood foundation sheets (which are the only type I have seen), there is a variation in cell size; some types being larger than natural worker comb although smaller than drone cell size. Plastic foundation of a larger cell size may not be interchanged with wax foundation of a smaller cell size. Some types of plastic foundation require a light coating of wax before use, while other makes can be used without any special preparation. It is not necessary to make a particularly tidy job of wax coating. Warm wax applied across the face of the frame with a paint brush is quite satisfactory; it only needs to stick to the hexagonal edges of the cell form and even then total coverage is not essential for the bees to make use of it.

Whatever type of foundation used, it is generally accepted practice to use worker cell-sized foundation for the brood chamber and drone cell-sized foundation for the supers. This is not just a fad: there are perfectly good reasons for this arrangement. Consider first the brood chamber. We wish to encourage the bees to produce as many workers as possible and only a few of the essential but non-productive drones. To achieve this, the brood chamber is fitted out only with worker-sized comb. The bees, not

Freshly built comb on plastic foundation. Note the corners are somewhat under-used; this is not unusual with this type of foundation *(author)*

to be outdone will produce some drone cells on worker cell-sized foundation by altering the cell size. This usually takes place along the bottom of those frames located near the outside of the brood area. As drone cells are larger and deeper, they are obvious even to the untrained observer. Furthermore, they are almost always clustered together producing a bulky and rather untidy effect. Where the drone cells are drawn out from the surrounding smaller worker cells some deformation takes place and unusable transition cells are formed at the junction of the two cell types.

Drone foundation is used exclusively for the supers and has several advantages over worker foundation for this particular application:

1. The larger and therefore less numerous drone cells give a greater area for storing honey.

2. Bees rarely, if ever, store pollen in drone cells while they will readily do so in worker cells. If worker foundation is fitted to a super immediately above the brood chamber, the bees will continue the band of pollen surrounding the brood area up into this super; however, if you are not concerned with show standard honey a little pollen will not do any harm.

3. In the natural state, bees often produce comb with drone-sized cells for storing surplus honey especially if there is a sudden glut. I have seen this in an empty super placed on a hive to hold a small feeder. The feeder hole was not completely blocked off and in the spring the bees built natural comb in the super. It was all used for storing honey and was almost all made with drone-sized cells.

Do not think you are confined to using shallow supers for storing honey. If you want to take standardisation to heart, you can use deep brood boxes and foundation for storing honey, but they will take some lifting if full.

Crownboards

Some confusion may arise by the manufacturers' referring to what is basically the same item as quilts, inner covers, clearer boards and crownboards. These subtle variations in description are not important. To all intents and purposes, the piece of wood that fits over the brood chamber, or supers if fitted, and forms a removable lid to the hive is a crownboard. In its simplest form it is a flat piece of plywood of a size compatible with the type of hive which it is intended to fit. In the centre is a hole or slot which will accept a bee escape or over which a small feeder may be fitted. If thin plywood $\frac{1}{4}$in (6mm) is used or bee space required over the frames then the outside is framed with strips of softwood $\frac{7}{8} \times \frac{1}{4}$in (20 × 6mm) thick.

Crownboards are available with one or two slots for bee escapes and return doors that allow the bees to bypass the bee escapes. Glass or Perspex panels may also be fitted. I am rather a fan of the glass or Perspex-panelled crownboards, especially for rapidly expanding small- or medium-sized colonies where some idea of their growth can be determined 'at a glance'. I only use them in the summer though; in autumn and spring they suffer from excessive condensation which is not healthy for the bees. Even in

From the top: undrawn foundation; drawn empty foundation; foundation filled with honey and capped (Jak Photography)

summer this type of crownboard exhibits some condensation, but I do not think this matters with the higher ambient temperature at that time of year.

If you decide to use only one crownboard per hive, then make it a wooden one with provision for at least one bee escape. When clearing supers of bees by using bee escapes, it is necessary to have another crownboard available, but this need only be a temporary item and one can save the expense of duplication of equipment.

Bee escapes

The bee escape is, effectively, a valve through which bees can pass one way but cannot return. In the early autumn it is necessary to remove the bees from the supers of hives to enable the beekeeper to take off the honey. A crownboard with one or more bee escapes fitted is inserted between the brood chamber and the supers. The bee escapes let the bees pass into the brood chamber, but prevent them from returning to the supers.

Possibly the most commonly used bee escape in Britain is the Porter, which has a central hole and two exits controlled by

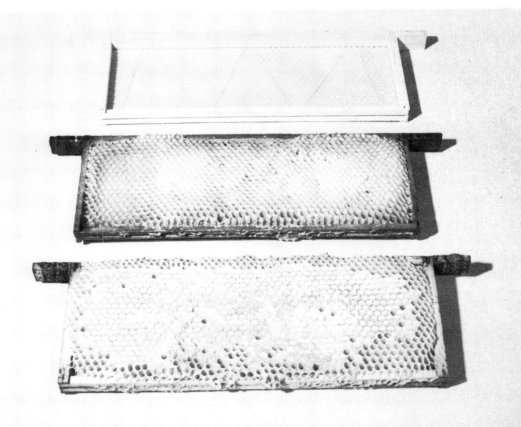

spring strips. It is very effective as long as the spring strips are adjusted correctly. Otherwise it will allow the bees to pass both ways. These are available in tinned steel and plastic, and can be dismantled, which is a great advantage when they are gummed up with propolis or if the spring strips need adjustment.

The Crowther is an eight-way bee escape which, it is claimed, clears supers quicker than the two-way Porter type. However, the Crowther seems to have declined in popularity in recent years, while a new arrival to the scene is the conical escape. This works rather in the manner of the funnel-shaped eel traps which are still used in the fenland waterways of East Anglia. The larger mouth of the cone is at the top adjacent to the super and the bees easily find their way into this and filter out through the small end and into the brood chamber. It is necessary to make up a special clearer board to use these conical escapes and up to five per board can be fitted.

So-called English and American pattern smokers. The large bent-nosed American type is preferable to the smaller straight-nosed English variety (*Jak Photography*)

Smokers

The smoker is one of those easily identified items of the beekeeper's craft. It is used to subdue the bees and works because of their instinctive fear of fire. Smoke is synonymous with fire, and in response to this danger signal they will gorge themselves with honey in preparation for possible evacuation of the hive. In such conditions, mildly panicked and sated with food, they offer little resistance to entry and examination of the hive.

There are two types of smoker in common use: the bent nose and the straight nose. The former is also called the American pattern and the latter the English pattern. The main component parts in both types are body, nozzle (or nose) and bellows. The body and nozzle assembly, which may be tinned steel, copper or stainless steel, is used to contain smouldering sacking or cardboard. The bellows, to which the body is attached, pumps air via a short tube into the body thereby forcing smoke out of the nozzle.

Most beekeepers favour the bent nose, American pattern smoker. It is larger and therefore holds more material and is more

stable when free standing. It is annoying to stand a smoker on an adjacent hive roof and then find it has fallen to the ground and is not to hand. This type of smoker can, because of its large diameter body, be refuelled without emptying it or losing the capability to smoke. When dealing with a large number of stocks this facility can be very useful. A further advantage over the straight nose or English pattern smoker is that it does not need to be inverted over the hive to apply smoke. The English pattern does and, because its narrow diameter body holds such a miserable amount of fuel, one can finish up low on fuel with a smoker blowing sparks, carbon and smoke into the hive. This is because the remaining sacking or cardboard falls into the nozzle as it is held over the hive and is pumped out along with the smoke.

Having started beekeeping with a straight nose smoker, I could appreciate the difference when I acquired a large secondhand bent nose type. Even for a couple of colonies I think it is worth having the larger smoker. It is very much a case of 'hang the expense, buy the best', for if you continue beekeeping that smoker could be with you for many years.

Queen excluder

The function of the queen excluder is to prevent the queen moving away from the brood area and into the supers. Even if the supers were furnished with worker foundation, we would not want the complication of brood and pollen mixed in with honey stores; but as it is normal practice to use drone foundation in the supers an uncontrolled egg laying spree by the queen would produce an abundance of drones. So, we exclude the queen (and incidentally the drones) by using a device that allows the passage of the worker bees but nothing bigger.

The queen excluder is produced in two basic types, the wood and wire and the slotted zinc or plastic type, and it is placed between the brood chamber and the supers. The dimensions must be suited to the type of hive used, so take care when purchasing queen excluders, especially at secondhand sales, to ensure that they are a suitable size for your type of hive.

The wood and wire type is a wire grid mounted in a wooden frame and qualifies as a 'rigid' excluder. Elderly examples are usually rather fragile and I would not recommend them if they can be avoided. A far better arrangement is the slotted zinc or plastic flexible queen excluder. When removing one of these from a hive, it can quite literally be peeled away from the propolis and brace comb which has glued it in place. As this type is essentially only a flat sheet with slots cut into it, cleaning is easy. It may be laid on a flat surface, such as a hive roof, and scraped clear of propolis and comb with a hive tool. I particularly like the plastic slotted flexible excluders as they do not deform like the zinc type and are much less prone to cracking across the bridges joining the slots.

Unfortunately, flexible queen excluders are only suitable for use with bottom bee space hives as they have to be mounted directly onto the top of the frames for support. However, they can be used for top bee space hives (normally fitted with rigid excluders) as long a some suitable method is used to support them. Strips of wood $\frac{7}{8} \times \frac{1}{4}$in $(20 \times 6mm)$ thick can be nailed or, with plastic, glued to the excluder to form an outer framework; it is also necessary to fix at least one support across the centre of the assembly. When used with a top bee space hive, such a modified excluder should be fitted with the wooden framework uppermost so that the correct bee space is preserved over the frames.

Hive tools

For my first two or three years of beekeeping I did not possess a hive tool. Instead, to lever off supers, part frames and carry out the other odd jobs normally performed with a hive tool, I used, at different times, a kitchen knife, an old chisel and even a screwdriver. It was not worth the inconvenience. My only advice on this matter is either to make a hive tool by copying one for a pattern or to purchase one. I do not think it is really even worth while making one unless you are so determined. Far better to buy the correctly designed item manufactured from suitable steel bar.

41

Feeders

Feeders are used primarily for feeding sugar syrup to the bees for winter stores. With a full-sized stock feeding replaces honey stores which have been harvested by the beekeeper. They may also be used for early feeding in the spring either to bolster meagre stores or stimulate the bees to embark on expansion of the colony. In an exceptionally poor year it may be necessary to use feeders at some time during the summer, usually with a large colony which has built up rapidly in the spring only to find a lack of natural forage available. Fortunately, this is a rare occurrence but it is nevertheless useful to have feeders to hand during most of the active part of the year.

I intend only to describe what may be regarded as top feeders, which are either placed over a hole in the crownboard or over the brood chamber complete in place of the crownboard. There are two basic variations: the contact feeder, which is the simplest and can be based on a tin can with a lever-off lid, and the rapid feeder, the best example of which is the Miller-type.

The rapid feeder is probably the most useful, although not necessarily the device which will be used by the absolute beginner. Rapid feeders may be made of wood, plastic, polythene or lacquered tin plate with a range of capacities from $1\frac{1}{2}$ pints to 2 gallons (0.8–9.0 litres). The large 2 gallon Miller-type feeder is built in the form of a wooden tray to cover the whole brood body. When fitted it resembles a shallow super and will hold sufficient sugar syrup for a complete winter's store. This type of feeder is preferable to the smaller rapid feeder which holds about 4 pints (2 litres) and requires several fillings to complete the winter feeding. The $1\frac{1}{2}$ pint (0.8 litre) feeders are virtually useless for winter feeding but are handy for feeding small nuclei earlier in the year. Whichever size of rapid feeder used (and the bigger the better) it is better to avoid, whenever possible, those made from lacquered tin plate; these are prone to rusting even when carefully cleaned after use.

The rapid feeder is not essential for the absolute beginner. Many beginners, and I was no exception, wish to start a hobby with a minimum capital outlay until they decide whether to commit themselves further or pack up and try something else. So, there is no need to purchase feeders when that common household item, the lever lid tin, is available. It can be converted into an efficient contact feeder by drilling twenty-five to thirty $\frac{1}{16}$in (1.5mm) holes in the lid. The tin is filled with sugar syrup, the lid fitted and the whole item inverted over a feeder hole in the crownboard.

The advantage of a contact feeder is that the bees do not have to leave the cluster as the syrup is available right over the top of the frames, while the rapid type requires the bees to move up into the feeder to fetch the syrup. Use of the contact feeder can be advantageous when the weather is cold and the bees not very active. A disadvantage, however, is that the bees take longer to draw off the sugar syrup than they would do for the same amount in a rapid feeder which has better access to the food. Another possible disadvantage which can easily be overlooked is self-emptying by expansion of the air pocket above the syrup. If, say, such a feeder is half empty after being placed on the hive the previous evening and there is a considerable rise in temperature the next morning, then the rapidly expanding pocket of air may force the syrup out of the holes in the feeder faster than the bees can draw it off. This is not normally a problem with the comparatively small feeders used, but if one were constructed to hold a couple of gallons then this venting of the sugar syrup into the brood area could be a problem.

Section racks

I would advise the beginner not to bother with sections; they really are a nuisance. You will have probably seen them on display at your local garden centre or health food shop. They are the square pieces of comb contained in their own little wooden box and presented in a fancy paper wrapping. Very pretty they look too, but to my mind a costly way of producing and selling honey. The basics for producing section honey are squares of thin wax foundation which are placed in wood or plastic (polypropylene)

An old WBC section rack. One sometimes comes across these still with part drawn sections in place, the wax usually riddled with larvae. It is quite unusable in this condition, and sections are generally more of a nuisance than they are worth (*Jak Photography*)

'sections'. These are mounted in a section rack with metal dividers placed in between each row for the purpose of ensuring that the bees draw out the comb evenly. The rack is placed on the hive in the same manner as a super. It should be noted that as section comb is made from worker-sized foundation it should not be placed next to the brood chamber otherwise some of the lower cells may be filled with pollen. Placing a section rack over a super avoids this problem.

If you wish to try a few sections without bothering to purchase or make a section rack then hanging section holders are a good substitute. These fit in a super in the same way that the normal frames are fitted. They are self-spaced and hold three sections each. The only proprietary ones available at present are manufactured to fit WBC or British National hives, but there is no reason why the principle should not be copied and adapted to other hives.

Round sections which have recently become popular in the United States utilise two-part plastic frames. I have not had any experience of using round sections and cannot comment on whether the bees use them more effectively than the older type of square section which they often seem rather loth to use. While bees will readily adapt to and utilise an undrawn super, they will often seem disinclined to use sections. It also needs a good honey flow for the sections to be drawn out, filled and capped as quickly as possible for then they will look their best. With a stop-start supply of nectar the sections will, over a few weeks, take on a rather yellowish, secondhand appearance, which is not an attractive selling point.

Protective clothing

There are two essential items of clothing: a combined hat and veil and a pair of gloves or gauntlets. Do not be fooled by pictures of

43

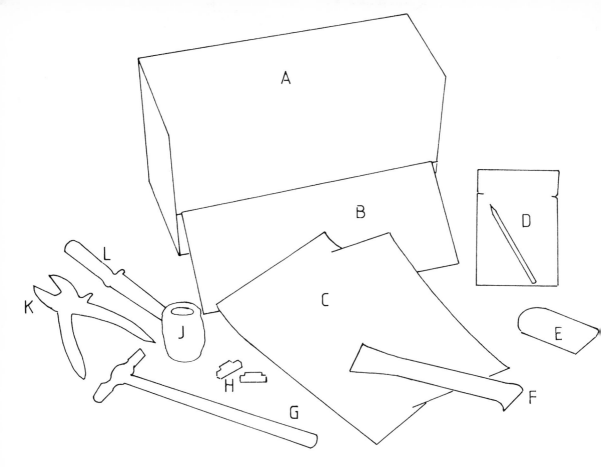

Fig 8 Key: (A) Box containing screw-top jar for collecting odd pieces of wax comb, and with a partitioned section for carrying two extra frames of wax. (B) The front of the box hinged open; the lid, hinged to the rear, not shown. (C) Blank hive record cards; used to keep track of the progress of a colony. It is useful to keep record cards under the roof of each hive, preferably in a polythene bag. (D) Note pad and pencil. A pencil is always preferable to a pen; it does not run out, is easily resharpened, and will write on wood if required. (E) Packet of panel pins. Always useful for odd repairs, especially if one has double-walled hives. (F) Hive tool (as described in text). (G) Small hammer to drive in panel pins. (H) Spare metal frame ends. (J) Ball of string. (K) Secateurs; useful when collecting swarms. (L) Knife; a decent open-bladed knife is often more useful than a small pocket knife

small children, wide-eyed and innocent, holding a branch with a swarm of bees attached apparently without the need for veil or any other protective clothing. They are posed pictures and the real world of beekeeping is not quite like that! It is essential that the beginner builds up his or her confidence even with the gentlest stocks of bees, and the best aid to this is to be well protected against bee stings. It is always necessary to use a veil and I find the combined hat and veil the most useful, although it is very much a matter of personal choice. If you have an old straw hat then purchase or make a separate veil. It does not matter how you arrive at a veil, as long as it gives good vision and is proof against the bees.

Soft leather gauntlets which are elasticated and fit almost up to the elbows are the best hand protection; household rubber gloves are a reasonable cheap substitute but can induce sweating on a hot day. Lined leather

It is always useful to carry a box containing the items shown. This particular box is constructed to carry two spare frames of foundation, as shown (*Jak Photography*)

Fig 9 The principles of (A) the contact feeder, and (B) the rapid feeder

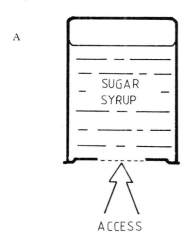

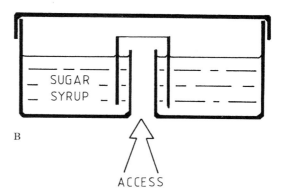

gloves are too thick and clumsy for carrying out sensitive operations.

For complete protection, overalls may be worn in conjunction with rubber wellington boots. Beekeeping overalls are usually white and there is a good reason for this: bees are least responsive to white or light colours. Do not be tempted to wear dark or navy blue overalls as bees have an extraordinary dislike of blue, and if the overalls are tainted with workshop smells they will often become very aggressive. In the absence of overalls, a pair of light trousers (not blue jeans) and a zip-fronted jacket are adequate wear. The jacket should preferably be elasticated at the waist and the trousers either tied at the bottom or secured with bicycle clips. If you have long boots or rubber wellingtons, then tuck your trousers into them. It is very disturbing to have a bee (or bees) crawling around inside your trouser leg as they almost always crawl upwards!

Whenever possible, when handling bees, avoid dark or woollen clothing or clothing contaminated with petrol or oil. Animal smells, perfume, after-shave and sweat odour are also disliked by bees and can inspire some aggressive reactions even from the gentlest of colonies.

Other useful items

The following items (some of them not normally associated with beekeeping) are also useful to have to hand:

1. *Travelling screen*: Manufactured after the pattern of the crownboard but with the centre section made from $\frac{1}{16}$in (1.5mm) perforated zinc sheet, it is used in place of a crownboard when moving the bees over long distances, and is an aid to good ventilation.

2. *Pollen trap*: A device to fit over the entrance of the hive which removes and catches pollen from incoming worker bees. The pollen is fed either to colonies lacking it or, particularly early in the spring, to aid the growth of a colony.

3. *Hive card*: A record card allocated to each colony. This is a very useful aid when dealing with more than two or three colonies.

4. *Moving board*: A temporary floor that can be used with a double-walled hive to convert it to the style of a box hive. Secured to the brood chamber with elastic straps, it makes a much more convenient package for transporting. The conventional floor, lifts and roof are then carried separately.

5. *Equipment box*: A homemade carrying box capable of holding everything except overalls and boots. An internal section constructed to carry a couple of frames is also an advantage.

6. *Sprayer*: A lightweight garden or household single-handed fine mist spray is useful for spraying a thin mixture of sugar syrup as an aid to some manipulations.

7. *Secateurs*: Ordinary garden secateurs are always handy, especially when swarm collecting.

8. *Stapler*: A lightweight stapler of the type used for fixing labels to wooden boxes can be useful for fixing a box hive together before moving it to another site.

5 Making a Start

So, you finally decide to have a go at keeping bees, but (let us suppose) you do not have any equipment, bees or contacts among beekeepers and the local beekeeping association is not listed in the telephone directory. Who do you contact? The most important first step is to get in touch with a local beekeeper, association or educational group. This can be achieved by contacting the British Beekeepers Association (see Chapter 1), who will be able to inform you of the name and address of the local beekeeping association county or area secretary. You could also contact the local education authority; they may know of a beekeeping evening study project in your area. If all else fails, when you are out for a drive in the car look for a roadside 'honey for sale' sign. Most beekeepers are happy to talk about their chosen hobby, especially if you explain your predicament—and remember to purchase a jar of honey!

Initial experience

Before any consideration is given to the acquisition of hives and bees, it is useful to have some experience of handling them to determine whether you are temperamentally suited to the task. I well remember the first hive I saw opened, and other beekeepers I have spoken to usually have a clear memory of their initial contact with what was then a strange new world. The first emotion is probably curiosity followed, as the operation progresses, with perhaps a tinge of panic. Bees are unlike other livestock; they cannot be led, cajoled or bullied. They respond to instinct and therefore do not have a fear of humans. The feeling of panic arises with the realisation that one needs to know how to handle bees or the situation can become out of control.

So, before beginning beekeeping yourself, try one of the following ways of getting experience. Attend a beekeeping evening study project, or offer to assist a local beekeeper for a season or part season. Such free assistance can be very rewarding, not only because of the knowledge and experience gained, but also for the chance of acquiring some equipment, albeit often rather dilapidated.

Obtaining equipment

At the outset of any new hobby an economical start with the minimum capital outlay is a prime consideration. With beekeeping some equipment can be obtained secondhand, other items must be purchased new, while hives and fittings can be a suitable project for the amateur woodworker. The internal fittings of the hive, such as the frames and wax foundation, should always be purchased new; dirty frames and old wax may be diseased and should not be considered under any circumstances.

New equipment and fittings do not represent a problem. Most manufacturers or stockists of equipment advertise regularly in various beekeeping journals and magazines, including *Bee Craft*, the official journal of the British Beekeepers Association, and the name of an agent operating in your area will probably be given. If not, write to the manufacturer of your choice, or preferably all of them, requesting a catalogue of their wares. Most operate a mail order service, although with the addition of carriage, packing and VAT it can seem expensive, especially if the order is small.

Many beekeeping magazines include private as well as trade advertisements. Some of these will be for bees and secondhand equipment. If there is nothing advertised in your region (and travelling to fetch bees is not an attractive proposition to the

It is always wise to examine a colony of bees prior to purchase *(author)*

beginner), then try the 'small-ads' in the local paper. If you do not find anything suitable, make use of the 'wanted' column.

Be patient, but keep trying and asking around. The neighbours' garden shed clearance up the road may unearth some useful long-forgotten piece of equipment. I have known it happen. Talk to the beekeepers you have contacted, go to displays and functions organised by the local association and make your needs known.

Choice of hive

If the intention is to purchase a secondhand hive, then the choice will be limited to the type of hive which becomes available. If cost is no object and includes the purchase price of a new hive, then the choice is more extensive. It is also useful when keeping a colony of bees to have a spare hive, so expenditure on this item can be doubled.

The two types of hive commonly used by amateurs in Britain are the WBC and National. The extensive use of the $14 \times 8\frac{1}{2}$in (35×21cm) long lug frame limits the logical choice to either of these hives, although the Smith hive using the same frame with short lugs is useful. If we therefore dismiss the use of the Modified Dadant, Commercial or Langstroth hive on frame size alone then the decision is restricted to the use of single- or double-walled hives. In spite of the theoretical difference between which is the warmest or has the best insulation properties, I do not think it matters where the choice lies, except in extreme climates. The following facts, however, should be considered:

1. WBC-type hives have the greater aesthetic appeal for a garden setting, especially when painted white.

2. National or Smith hives (box type) may not be so attractive as a garden ornament but are easier to handle, especially if they have to be moved.

3. In my experience, the WBC hive can be drier in a very wet winter but is also a favourite nesting site for mice. A WBC hive in need of repair is often a better proposition than a box hive in a similar condition; woodworking accuracy is not so important so it is more suitable for home repair.

4. The National hive has the advantage of taking eleven brood frames against the ten of the WBC; most reasonable colonies of bees will need eleven frames.

5. WBC hives with their lifts, legs, extended flight boards and entrance sliders can be a nuisance because of the abundance of parts; there are so many variations in the size of lifts that standardisation of secondhand equipment is an enormous problem.

6. Whatever the final choice, consider possible future expansion and then the need for standardised equipment. It is this very real need for standardisation that makes the box hive such an attractive proposition. I prefer the National hive because of its fairly extensive use which ensures a good supply of secondhand spares and accessories. While such hives usually cost more on the secondhand market, they are cheaper to purchase new than the WBC.

Stocking the hive

Despite the disadvantage of having 'all one's eggs in one basket', I think it is best to start beekeeping with one colony of bees. If your experience is very limited then a nucleus colony is preferable as it is small and easy to handle. The possible drawback with a small or part colony is that it gives less room for error should anything go wrong, eg the loss of the queen. Nor will there be any surplus honey from such a colony unless it is formed early in the year, say mid-May, and even then it could be a poor season not conducive to growth or honey production. However, an overwintered four- or five-frame colony will usually increase sufficiently during the following spring and summer to give a modest surplus of honey.

There are several ways of obtaining your first colony of bees: (a) by the purchase of a nucleus colony; (b) by the purchase of a full colony; (c) by purchasing or finding a swarm; or (d) by the gift of a colony, or a swarm.

Nucleus colony
Normally a nucleus colony consists of four or five full frames of brood and emerging brood, well covered with bees and complete with queen and the equivalent of a full frame of food. It is possible to build up a colony starting with only two full frames of brood. However, to enable them to build comb and store sufficient food for wintering such a small nucleus should not be considered after the end of June.

If a nucleus is purchased from a beekeeper in the locality, it has the advantage that it is formed from bees native to the area, which may prove hardier than a strain imported from further afield, but take the following precautions:

The basic operations involved in opening a hive—in this case, for examination of the honey stored in the supers (*author*)

(*opposite top left*) Ready to go—stand behind the hive to avoid the flight line of the bees

(*opposite top right*) Remove the roof

(*opposite bottom left*) Apply a little smoke under the crownboard; this helps to drive the bees down out of the supers

(*opposite bottom right*) A little more smoke may be applied through the hive. This is something that can be judged with experience

(*left*) Remove the crownboard

(*below*) Drawn out super frames containing honey, but as yet uncapped

1. Ask to see the hive opened.

2. Examine the condition of the wax and frames. Some beekeepers make it a policy to replace the oldest frames in the hive with new wax and use the old frames to form nuclei. This is not necessarily a bad thing but very blackened wax and frames stained dark with propolis are an indication that this may have been done.

3. Look for approximately a 20 : 20 : 60 proportion of eggs, grubs and sealed brood in the centre frames with some pollen evident and honey crammed into the corners. The outside frames should contain a high proportion of honey and pollen. If the outer halves of the outside frames are exclusively composed of stores this is all for the good.

4. Check for any signs of disease, although with limited experience this is difficult. I well remember finding a filthy looking substance in some brood frames and after several worried hours discovered that it was stored pollen. Most beekeeprs seem to be fairly honest, especially if they make a practice of selling a few nuclei each year, and there is a greater chance of finding disease in a neglected colony or swarm. If the colony appears active with flying bees carrying pollen and plenty of brood, then there is a good chance that your purchase will prove satisfactory.

5. You can ask to see the queen but, as a novice, it will have little significance except for the satisfaction of knowing there is one present. The evidence of her activity in the visible form of eggs laid in the cells is more important.

6. It is always worth noting the condition of the vendor's equipment if it is in the vicinity. An immaculate turnout of hives standing in regimented order is not necessary, but if they appear rotten, dilapidated and surrounded by undergrowth it may be best to go elsewhere.

Full colony

Examination of a full colony is little different from the examination of a nucleus. The basic checks for disease, the condition of frames and so on are all applicable. If it is a warm midday when most of the flying bees will be absent, there should still be sufficient bees in the hive to cover all frames. A colony in midsummer with a strong laying queen will often have eggs and brood to the inner face of the outside frames. There should also be honey stored along the top edges of the frames in extra comb.

Swarms

Unless they are from a colony with known characteristics, swarms can be a gamble. Collecting a swarm, depending upon its location can be easy or may be very difficult, and it is preferable to have had some previous experience and have a competent associate on hand.

A swarm with a known ancestry can prove very useful. Some 'nuclei' offered for sale are made up from such swarms, hived quite often on a few drawn frames. The swarm may contain an older queen but this is not necessarily a disadvantage. Stray swarms, however, are an unknown quantity and, until they are settled in a hive, do not show their true temperament. There is always the possibility that they may carry disease and one must always be prepared to accept the possible loss of a swarm, the frames on which it has been hived and the cost of feeding. However, a swarm is, at least, remarkably durable, and when hived will draw out comb and become established in a few days.

It should be possible to obtain a swarm for free or, at the most, only a few pounds should change hands. If you have difficulty obtaining a swarm through lack of contacts, then advertise in a local paper. Advertising for swarms, especially if there is a hint of a reward, really does work effectively. Most people regard swarms as a nuisance or possible danger and are usually very willing to contact someone who is prepared to remove them. Inform your local police station that you may be contacted when some worried person reports a swarm, but be prepared to accept that it may be lodged under the Town Hall balcony or some other equally inaccessible place!

If you do decide to advertise for a swarm, be prepared for a few wasted journeys. There are nearly two hundred species of bee in Britain and most people do not know what the honey bee looks like. I have been asked to remove so-called swarms that have included 'bumble' bee nests and, on one occasion, a number of bees harmlessly sunning themselves on a gravestone.

Gifts of a colony or swarm
If you are fortunate to be given a hive of bees or a swarm then consider yourself lucky, but if you receive a full colony do not expect to find it necessarily in the best condition. It may well be perfectly healthy but well established, ie stuck together with propolis and brace comb with perhaps some natural comb. The full complement of frames may not be present, or the frames may not be aligned correctly. Never mind, it is all useful experience and not difficult to cope with if tackled methodically. Methods of dealing with such minor problems are outlined in Chapter 7.

The bare essentials

If you are considering keeping only one colony of bees initially, possibly starting from a nucleus, there is a certain minimum amount of equipment necessary simply to hive the colony and allow for growth. If a full colony, five-frame nucleus or a larger overwintered colony is acquired early in the year, say May, then allowance must be made for honey production. The following lists of equipment cover both situations. They are compiled on the equipment requirements for a National hive, although the principle is the same for any box hive. If a WBC hive is used, remember to add the required number of lifts.

Minimum equipment
Hive
  Brood chamber
  Floor
  Entrance block
  Crownboard
  Roof
  Division board

Internal fittings
  British Standard deep (brood) frames (Hoffman self-spacing frames preferred) sufficient to add to nucleus to make up full number of frames
  British Standard brood, wax foundation Wired or plastic foundation sufficient for frames obtained

Accessories
  Veil
  Gloves
  Smoker
  Feeder

Additional equipment
If the nucleus is large or a full colony has been purchased, the following additional equipment is required.
Supers
  2 supers
  1 queen excluder
  18 British Standard shallow frames
  18 sheets British Standard shallow wax foundation (drone). (Wired foundation should be purchased if centrifugal extraction is considered, and specially thin [non-wired] wax foundation can be purchased for cut comb)
  1 spare crownboard with provision for bee escape(s)
  Bee escape(s) to suit crownboard

Siting the hive

If one or two hives are to be kept in the garden and it is a small garden bounded by neighbours, then some care is necessary in siting them. It is helpful to mention your proposals to any immediate neighbours, not forgetting to explain the benefits of pollination and hinting, perhaps, at a jar or two of honey in the future. In my experience, most people are not too worried about bees based in someone else's garden; after all, they are used to seeing them flying around the flower bed without offering any harm. At the same time, it is surprising how many of the wasp stings inflicted on children in the area will be attributed to your bees. If there is strong objection from the neighbours and you do not wish to antagonise them, then look for a site close to home, preferably within easy walking distance.

When siting a hive or hives in the garden the following aspects should be considered:

1. The hive(s) should be facing away from any paths in the vicinity. Bees flying from the hive tend to fly fairly low and do seem to object to human or animal obstruction.
2. For the same reason the flight line of the

The bare essentials *(Jak Photography)*

bees coming from and going to the hive(s) should be as high as possible. This can be achieved in part by facing the hives into a wall, fence or hedge. The bees returning to the hive(s) will therefore fly in over them, ensuring that they are approaching from a reasonable height.

3. The chosen site should be some distance from the house, say a minimum of 10yd (9m), for those occasions when the hive(s) are opened when there will be a considerable number of insects milling around.

Sometimes these requirements cannot be accomplished with natural screening, so erect an artificial barrier. Fine nylon netting on a simple frame is most suitable. It is not necessary to cage the hive(s) on all sides unless it is intended to keep the family dog at bay. Bees dislike dogs and will attack them quite savagely if they should venture to sniff the front of the hive or indulge in canine territorial marking.

It was generally held at one time that hives should always be faced south or south-east. This need not be adhered to strictly, but it is advantageous to find a sheltered area with a southerly aspect. For instance, if hives are being kept adjacent to buildings it is best to place them on the south side but face them whichever way is most convenient. Avoid the north side of a building, which can be quite dismal and damp in the spring when the bees need some warmth to stimulate them. Likewise, damp hollows at the bottom of a hillside or next to a stream or ditch should be shunned. Trees can provide excellent shelter from the midday sun but a heavy canopy of branches directly over the hives only provides a damp gloomy environment which again is unsuitable. A clearing in a wood is quite different. Such an area is often warm and sheltered and has the advantage of being hidden from the road and from vandals.

When siting an 'out apiary', an apiary situated away from home, take into account the presence of large domestic animals. Cattle and pigs are somewhat insensitive to bee stings and may use a hive as a rubbing

54

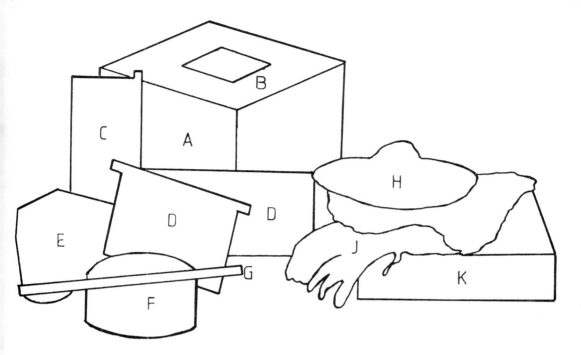

Fig 10 Key: (A) Brood box and floor. This British National hive will accept eleven frames such as the Hoffman type. (B) Crownboard fitted with glass panels. A piece of plywood is shown placed over the feeder hole while not in use. (C) Division board; if less than the full complement of frames is fitted, this board is used to prevent access to the rest of the hive. (D) New Hoffman type frames containing undrawn foundation. These would be placed either side of a nucleus colony to accommodate growth. (E) Smoker, the so-called 'bent-nosed', American or Root's type. This particular one has the nozzle and firebox made of tinned steel; copper or stainless steel is preferable. (F) Rapid feeder; this is made of lacquered tin plate, and holds about 4 pints (2 litres) of sugar syrup. (G) Entrance block to control the size of the hive entrance. (H) Hat and veil. This type of combined hat and veil is very popular and convenient. It is held in place, when worn, by elasticated straps under the arms. (J) Soft leather gloves fitted with twill gauntlets. (K) Hive roof, shown inverted. When transporting a hive it is placed inside the roof

hives or using a catapult for greater effect. Sometimes hives are deliberately knocked over and this can result in the loss of the whole colony. There is little that can be done except to make the hives as inconspicuous as possible. Base them preferably within sight of habitation but as far from a main road or public right of way as is practicable, remembering that you will require vehicular access yourself. There is little advantage in having hives securely placed if they are three fields away from the nearest farm track, especially when removing the honey-filled supers or on the occasion when the smoker is inadvertently left in the car.

post. Bee hives do not stand up to this sort of treatment and can easily be pushed over and trampled to destruction. If there is any likelihood of farm stock coming in contact with the hives, either obtain permission to fence them off or choose a different site.

Unfortunately, fencing around hives will do little to deter vandalism. Usually this is confined to throwing stones or earth at the

Handling bees

I was once asked at a honey show whether I talked to my bees. I replied 'Only when they sting me' and at such times it is usually not repeatable! Talk to your bees if you must but it will not make any difference except as an expression of your own feelings. It is better by far to study your insect charges quietly and find out how they react to being handled at certain times of day and under different weather conditions.

Sufficient equipment to house a full colony of bees, and hopefully produce two full supers of honey. Note: not visible is the plastic short slot queen excluder fitted between the brood box and the super *(Jak Photography)*

There are several factors that affect the response of a colony under examination: the inherent temperament of the colony, the weather, the time of day, the time of year, the size of the colony, the presence of disease and the prevalence of robbing. The temperament of a colony can only be altered by requeening with a queen from a strain noted for certain characteristics. An average colony will tolerate examination of the hive with little or no resistance under ideal conditions, but may react somewhat aggressively under less than perfect conditions. It will exhibit a non-aggressive response when exposed to the following conditions: early in the season; on a warm sunny day; when the colony is small; when it is a newly hived swarm; when it is weakened, queenless or diseased. An aggressive response will be exhibited: late in the season, especially when supered; in the late evening; in thundery weather or when

rain is imminent; when robbing is prevalent in the apiary.

The way in which a hive is opened and examined is important. Do not stand in front of the hive or directly in the flightline. Handle the frames gently; do not wrench them free or knock the sides of the hive and try not to squash or crush bees as you withdraw frames. If a division board is fitted to the hive, remove it first to give you more room to manipulate. If it is a hive not designed to take a division board, then remove one of the outside frames first. It can be placed on the upturned roof or, if using a WBC hive, hung by its lugs diagonally across a lift. It is easy to make up a couple of wire lugs in a flattened 'S' shape to hang a frame on the outside of a hive and proprietary items are available for this purpose. One note of caution: check that the queen is not present on a detached frame. We do not want to chance losing her. It is unlikely that she will be on the very outside frame but is nevertheless a possibility not to be ignored. You can always shake the frame over the hive with a sharp downward movement to rid it of

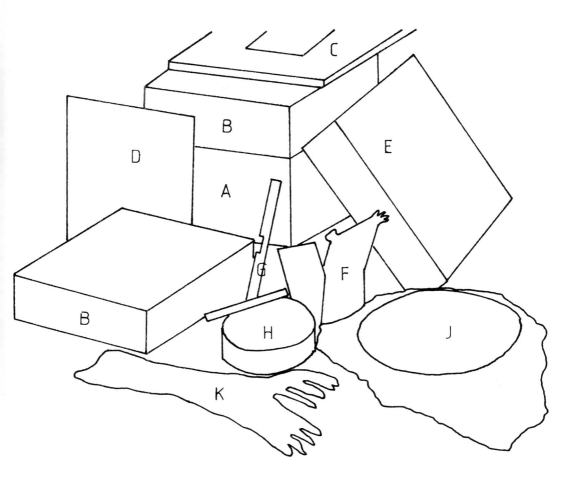

Fig 11 Key: (A) Brood box and floor. This British National hive will accept eleven deep brood frames. (B) Supers containing nine shallow super frames. Both supers contain drawn out foundation that has previously been extracted. The super in front of the hive is fitted with castellated spacers for frame spacing. (C) Crownboard with glass panels. The feeder hole is blocked with a piece of plywood to prevent bees leaving through the crownboard. (D) Spare crownboard fitted with Porter bee escape. This is fitted over the brood chamber when it is necessary to clear the bees from the supers. (E) Roof; note the zinc cover. (F) Smoker. A favourite trick to put the fire out in a smoker is to push a wad of grass into the nozzle as shown. The best fuel is old hessian sacking bound into a bundle, which is then lit and allowed to smoulder steadily, and then put into the smoker. (G) Entrance block to fit hive. (H) Rapid feeder. A useful sized feeder, but it will take three to four fillings to supply sufficient winter stores. (J) Combined hat and veil. (K) Soft leather gloves fitted with twill gauntlets

most of the bees. When examining a frame of comb, stand holding the frame over the brood chamber so that any bees dropping off land in the hive rather than have them milling about on the ground or crawling up your trouser leg.

Early in the season or when a colony is small, few bees will waste their efforts attempting to repel the intruding beekeeper. That same colony later in the year with stored honey and an abundance of flying bees can prove quite a handful. It is the flying or foraging bees that are inclined to attack and therefore it is best to avoid examination in the late evening when many of these bees will be back at the hive. During the day, impending changes in the weather are likely to be the most noticeable factor affecting the bees. Most colonies, even the gentlest, do not like rain, especially thundery weather. At

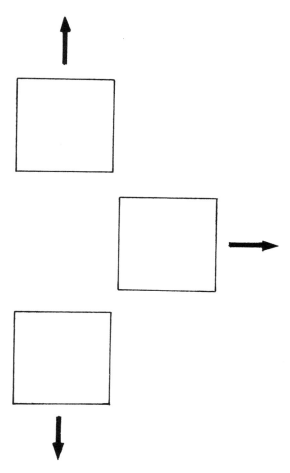

Fig 12 A simple three hive layout, with the entrances at different angles. This is also a very convenient set-up to work with

The smoker is a comforting device to have to hand, but do not overdo the use of it. Some bees can be handled with little or no smoke, while others if smoked excessively get very excited and will pour out of the front of the hive. A couple of puffs at the entrance, and perhaps the same again under the crownboard, will suffice for most colonies, with an occasional puff of smoke drifted across the top of the frames to keep them in order. Do not bother directing smoke at flying bees buzzing around your face; they are only curious. Likewise, do not flap your hands and try to drive them away: it is a waste of time. It is difficult for a beginner to accept that the majority of bees milling around mean no harm, but try standing still and watching them. Many will alight on your arms or legs, perhaps on the veil, will wander around for a few seconds and then lift off again having lost interest.

A bee intent on stinging will attack without any preliminaries, directing its efforts against those parts of your anatomy nearest to the hive and even the smoker on occasions. They will instinctively go for your face and eyes, and when quite frenzied will fly headlong into your veil, which is the very best of reasons for always wearing one. If you are ever in the unfortunate position of having a really savage colony then consider starting your first requeening experiments.

A really savage colony is hard work and no fun at all. I had one colony which was a hived swarm found clustered on the ground in an orchard. For the first few days in their new home, as they established themselves, they were quite docile. After about a fortnight the fun began. They would attack anything within a few yards of the hive. Lifting the crownboard resulted in a roar of aggression with which they would pour forth and attack in quite alarming numbers. I wrapped up well and only got stung a few times, but working with twenty or more bees scrabbling around the veil trying to get in is quite disconcerting. Eventually I destroyed the queen, and after a few days united the colony with another which had a much quieter natured queen.

Fortunately, such aggressive bees are not common (perhaps less common than those which are particularly easy to handle), but

such times their response can be so savage that it is best to retire and leave them until later. This should not be regarded as an admission of defeat and is nothing to be ashamed of. Many experienced beekeepers will tell you that occasionally it is best to close the hive and walk away.

It is useful when gaining experience of bees and beekeeping to examine colonies under the most ideal conditions. That way the budding beekeeper can build on his or her experience and confidence. Later, when a little more experienced, awkward operations may be carried out under less than ideal conditions; then if anything goes wrong it should not dent your confidence too much.

they are more noticeable. It is the one occasion when a heavy smoking can be useful. Pump some smoke into the entrance and leave them for a few minutes before operating on the hive in the normal manner. They should be a little easier to handle.

Preparing the hive

If you have made or purchased a new hive then, apart from assembly, preparation of the hive to accept a full colony, nucleus or swarm is less time-consuming than preparing secondhand equipment.

The new equipment will first have to be assembled. Most British manufacturers will supply hives 'in the flat'. In other words, in component form with the requisite number of nails, tacks and fitments to build it complete. This method of purchase aids transportation and is subject to a financial discount. Instructions on assembly are almost always supplied, but some are more explicit than others. There is normally little difficulty in following these instructions and it is worth while taking one's time to ensure a first rate job. After all, the completed hive may be standing outside for many years with only the occasional coating of preservative to protect it from the weather. When the hive is completed, the exposed wood on the roof, body, entrance block and floor should be painted with some form of preservative.

The frames of foundation can, like a hive, be purchased ready assembled or in the flat. They are so simple to erect that there is really little reason for purchasing them already assembled. If a full colony is being housed in the hive, then it is useful to have a few extra brood frames of foundation ready, either to hive a swarm or perhaps start a nucleus. In many cases the beginner will be preparing his or her hive for a part stock in the form of a nucleus or swarm. In such cases a division board should be fitted.

A proper division board will, when fitted against an exposed end frame, seal off the unoccupied portion of the hive. This is essential when hiving a swarm of bees which may ignore the frames provided and build their own natural comb on the wrong side of the division board. As long as a part colony of bees already on the frames has sufficient

room for expansion in the form of empty frames then there is not such a problem. A useful makeshift division board can be constructed by tacking a thin sheet of aluminium or plastic of suitable dimensions to an empty frame.

The only other item necessary to complete the brood chamber assembly is a crownboard, which when in place should have any holes made to accept a feeder (if not fitted) or bee escapes blanked off. A piece of $\frac{1}{4}$in (5mm) plywood of a suitable size can be placed over the holes; the bees cannot move it and will concentrate instead on securing such loose items with propolis. It is useful to have a number of such pieces of wood available in the apiary for these purposes. The best place to keep them is ready to hand under the roof of a hive.

Secondhand equipment has the advantage of not needing assembly but may need repair and will certainly require treating for possible traces of disease. A simple and effective method of eradicating any spores of disease is to scorch the inside of the brood chamber, the floor, the underside of the crownboard and the division board using a gas torch or even a twist of burning straw. When this has been done, it is necessary to wire brush both the inside and the outside of the hive thoroughly. The outside of the hive can then be painted or treated with a suitable preservative.

I never use secondhand frames of unknown origin, although it can be argued that treatment with acetic acid will effectively sterilise them. I would rather err on the side of caution than chance introducing disease into the apiary.

During recent years there has been an increase in the number of hives stolen from out apiaries. To act as a deterrent, some beekeepers brand their hives for identification. Several suppliers of beekeeping equipment list branding irons which may be a combination of letters or a unique design. The iron should be heated to a cherry red and burnt well into the wood so that it cannot be easily sanded off or obliterated. All main sections of a hive should be so marked.

These old and very obviously non-standard hives may not look much, but are usually very economic purchases, and with a little time and patience can be made into a sound home for a colony of bees *(by kind permission of Gloucestershire College of Agriculture)*

Water supply

During warm weather, especially at the peak of the breeding season, a keen observer may notice bees drinking from a pond or at the edge of a stream. This is not unusual: they will drink from almost any readily available water supply.

There is a pond in my present garden and a small apiary within a hundred yards. A regular traffic of bees comes to the pond and sometimes a dozen or so can be seen drinking at any one time. They soon learn the best places to drink from: where the edges slope gently away and there is little danger of the bee falling in the water.

It is possible to construct a simple water catchment on an out apiary site. A favourite method is a couple of old corrugated roofing sheets sloping into a suitable receptacle. Care must be taken that the sides of this mini reservoir are not too steep or slippery otherwise bees can fall in and drown. This can

be avoided by floating a flat piece of wood on the surface which they will quickly learn to use.

In most instances I do not think it is necessary to construct such a receptacle, particularly in Britain where water is not normally scarce. We may not notice it, but there is water everywhere: ponds, streams, gutters, ditches, even large puddles. Anyone who has ever flown over central London will confirm that even in this large city there are surprisingly large tracts of water, and the growth of interest in the ornamental garden pond ensures that there is a more than adequate water supply even in the suburbs.

Secondhand equipment

Most of us, at some time or other, purchase or are given secondhand equipment which may not be in the best condition. A deliberate purchase would tend to indicate that the equipment was a much-needed addition, while gifts of old bits and pieces may range from the useful to the utterly useless. Never refuse any such gifts, even if it comprises two or three homemade WBC supers riddled with woodworm. You can always dispose of

60

real junk afterwards, but if you develop a reputation for gracefully accepting the odd piece of equipment then sometime in the future a real gem may turn up. Do not be mean. As well as thanking a donor of equipment, part with the occasional jar of honey; it could be a worth-while investment.

When I first started beekeeping, a surprising number of people in the village dug out long-disused pieces of equipment. At one time or another I have been given broken queen excluders, two English pattern smokers with rusty bodies and cracked bellows, a badly damaged WBC hive without any internal fittings and a quite rotten Double Conqueror hive filled with old comb infested with wax moth. All such gifts have been gratefully received because at other times I have been the fortunate recipient of nearly a dozen good queen excluders, five sound British National hives, a hive tool, a small extractor only in need of a coat of paint and one excellent colony of bees. Diplomacy

can pay, sometimes almost as well as the return from keeping bees.

Even if you are unfortunate enough not to be given any equipment, it is still quite likely that some equipment will be purchased secondhand. The most common secondhand purchases seem to be hives, supers, feeders and honey extractors. I will deal with hives shortly in greater detail. Feeders found at sales are often the small tinned steel type and of no great use. Similarly, honey extractors of the tinned type may be either rusty or coated with old paint. As long as they are sound this does not matter. A good clean up and a fresh coat of paint will make all the difference. With a steel-bodied extractor make sure it has not rotted through where the sloping base joins the sides. This is the normal problem area when honey has not been cleaned out properly from previous use. Items such as crownboards, the useful Miller-type feeders, settling tanks and wax extractors seem less common commodities on the secondhand market, and most decent extractors are snapped up fairly quickly.

Equipment such as feeders, settling tanks or ripeners and honey extractors made from

A simple homemade branding iron for marking hives in an attempt to deter theft (*Jak Photography*)

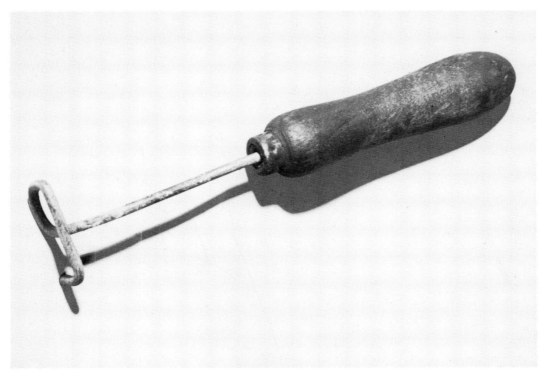

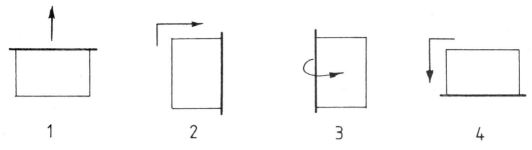

1 2 3 4

Fig 13 Handling frames

Photographic sequence (opposite) Handling frames. This method of handling frames for examination originated with the old unwired foundation, and is probably still the best method, especially when there is uncapped honey in the comb. Although these posed photographs make the operation look a little clumsy, it becomes much easier with practice *(author)*
(top left) Withdraw the frame in the normal manner
(top right) Turn through 90
(bottom left) Swivel the frame through 180
(bottom right) Turn the frame through 90° so that the top is now at the bottom. This enables one to examine both sides of the frame without the risk of the unwired comb falling out. It is also useful when there is stored honey in uncapped cells, as it prevents it running out. To put the frame back in the hive, simply reverse this sequence

(below) Removing a zinc short slot queen excluder from a modified National hive. There are several interesting features: the entrance block is missing from this hive; the

tinned sheet steel should be examined for soundness if they look rusty. A light deposit of surface rust is of little consequence and a shabby paint finish is nothing to worry about. All such pieces of equipment can be cleaned down if necessary and painted with polyurethane paint. Two coats of paint is usually sufficient and will prove to be much more durable than the original tinplate. All painted items should be washed out well with hot soapy water, followed by a cold rinse, before use. The same items made of aluminium, stainless steel or polythene will not require such attention and this is usually reflected by the higher prices they command in a sale.

stand is an inverted plastic milk crate; the smoker can be conveniently held between the knees while working on the hive *(author)*

Double-walled hives are usually constructed of flimsier material than box hives. The outer lifts in particular will often display signs of loose jointing but reinforcing can be added if required. The biggest enemy of all hives is wet rot and, because of the nature of its construction, the double-walled hive suffers most of all. Areas to examine closely are:

1. *The legs*: Check that they are free from rot and that the hive stands evenly. It is not uncommon to find a WBC hive with odd legs.

2. *The flight board*: Wet rot may occur where the flight board fits up against the bottom lift. Quite often the whole board may be in poor condition, but as long as the supports underneath are sound it is easily replaced.

3. *The porch*: This is always a problem area, wet rot occurs around the top edge where the water runs down the main body. It may also be insecurely attached, and note whether the guides are fitted for the sliding shutters.

4. *The bottom lift*: It is the bottom lift, to which the porch is fitted, that is exposed to the most weathering and particular attention should be paid to this whole assembly.

5. *The roof*: Check for rotten or broken boards in the roof. If old polythene fertiliser sacks have been nailed on as a waterproof cover, the woodwork may well be rotten. See if the bee escapes are still attached; but if not it is of no great consequence as they are cheap enough to replace.

A general check should be made for the presence of woodworm in any wooden hive. This can be eradicated by a good soaking with paraffin or creosote. Unfortunately, the use of creosote means that the equipment will not be ready for use for some time.

The fit of supers together, and to brood boxes, should be carefully examined. Look for a twisted assembly or gaps in the jointing. Sometimes homemade supers intended for double-walled hives are constructed with the most appalling mismatching, a situation that cannot be tolerated with box hives.

Box hives are much more durable, being constructed of stouter timber than double-walled hives. A check should be carried out for loose or rotten corner joints, and the National hive in its modified form can exhibit problems when water has seeped between the lower locking bars and the inner walls of the hive.

6 Spring

Spring does not start officially until 21 March, but once Christmas is over I find myself eagerly looking forward to examining the contents of each hive, although this is often tempered with apprehension, wondering how the bees have fared through the winter. I still find it incredible that they survive in what are little more than flimsy wooden boxes. A group of hives huddled in the lee of a barn or against a bare hawthorn hedge seems so lifeless and empty, like coffins. I always feel better when a few bees are to be seen around the hive entrance on a sunny day. Sometimes even in February, if it is exceptionally warm for the time of year, a large number of bees will take flight and for a brief period the activity is reminiscent of the more hectic days of summer.

Winter's grip, however, is slow to relax and doubts soon flood back about the bees' welfare. Was the sugar syrup fed early enough in the autumn to avoid fermentation of the stores? Perhaps it was a mild autumn, so do they now have sufficient food left? Has the queen survived? Without her this early in the year the colony is doomed. Most of these questions must remain unanswered until a little later in the year.

Spring cleaning

During the winter months as the bees slowly consume the stores, discarded cappings, old pollen and dead bees fall to the hive floor. A strong healthy colony will clean out much of this debris on a sunny day but a weaker colony may not. If the weather has been cold for some time the bees will not have been very active and the rubbish will accumulate. It can become quite a filthy mess and is best cleaned from the hive floor. For those of us afflicted with the impatience of the beginner, it is an ideal excuse to do something with the bees after the winter layoff.

It is a job best carried out in late February or early March. Choose a bright, cold day when the bees will be tightly clustered and disinclined to fly. If the operation is carried out gently in such conditions, the bees will hardly be disturbed.

It is useful to possess a few spare clean floors and simply exchange them for those on the hives, but if you do not have any spare floors then each hive floor must be cleaned and replaced. It is a simple job with box hives and only a little more complicated with double-walled hives. The following is a simple breakdown of the sequence of operations required if using a box hive. The brood chamber, frames and crownboard assembly are referred to here as the body of the hive.

1. Remove the roof and stand it upside down to one side of the hive.
2. Gently prise the body of the hive away from the floor with a hive tool. (Note: the floor is usually glued on with propolis and it may be difficult to part from the body of the hive without jarring the contents.)
3. Place the body of the hive crossways on the inverted roof.
4. Remove the entrance block, scrape the floor clean with the hive tool or replace the floor with a clean one. Replace the entrance block.
5. Place the body of the hive back on the floor making sure it is aligned correctly.
6. Replace the roof.

With a double-walled hive of the WBC type the sequence is almost the same except that the outside lifts must be removed and the body of the hive placed across a lift rather than the inverted roof, especially if a gabled roof is fitted.

There are a number of incidental but none the less important pieces of information to be gleaned from this early 'spring cleaning'. It is

The consequence of disease and neglect: piles of dead bees and a mouse nest. The natural comb shown on the lower right hand frame was caused by excess space between the frames. The resident mouse ate through some of the other foundation *(Jak Photography)*

useful, before scraping the floor clean, to examine the deposited rubbish. Given the right climatic conditions, a strong colony will have little mess on the floor; often only a few cappings from the emptied honey stores, perhaps some traces of pollen and a few dead bees. If the weather has been particularly cold for a number of weeks preceding the examination, you may find a larger number of dead bees, perhaps some twenty or thirty. This is nothing to worry about. When the hive floor is black with dead bees there is much more cause for concern: either they are starved of food or weakened with disease.

Spring feeding

There are two reasons for feeding during spring: either to provide food for a colony near to starvation or to stimulate expansion of the colony. Early in the year we are only concerned with replacing depleted stores.

Feed the bees early in the spring only if you have determined that they may be short of stores, otherwise it is best to leave them undisturbed. If you do need to feed them then either candy or sugar syrup can be used. I am still not sure what is the best food to provide this early in the year. Candy, it will be pointed out by many beekeepers, is an unnatural food, but to me it seems no different from finely granulated honey, such as that obtained from oil seed rape. The bees will consume it whether they are starving or not. A friend of mine uses candy as a kind of insurance against 'local starvation'. This is a situation where there may be stores available in the hive but the colony becomes separated from the food and consequently starves. This is only likely to happen if it is very cold for some time and even then is an uncommon occurrence. If the bees come up against a block of candy it gives them a food supply until they can locate the rest of the stores.

Sugar syrup as emergency feeding is best fed thick and warm in a contact feeder rather than a rapid feeder. The use of a contact feeder avoids the necessity for the bees to leave the cluster which they need to do with a rapid feeder. It can often be difficult to entice bees up into a rapid feeder when the weather is cold, but a contact feeder makes the food instantly available.

On those occasions when I have found bees in need of food early in the year I have fed a full colony about 5lb (2.25kg) of sugar made into a thick syrup of the same consistency as the autumn feed (see Chapter 8). Leaving them for about a week, I replace the feeder (usually a tin with holes in the lid) with a 1lb (450g) block of candy. It is no use being restrictive with feeding, and if that means feeding autumn and spring to ensure their survival then so be it.

Cleansing flights

Under normal healthy conditions bees do not defecate in the hive but take a short flight to empty their bowels. That is not to say that they need to be in flight to carry out this simple function. A light-coloured vertical or near vertical surface reflecting the light and warmth from the sun often attracts their attention. Evidence is left in the form of what look remarkably like miniature bird droppings.

If it has been cold for some weeks, followed by a day or two of sunshine, quite a large number of bees will leave the hive on cleansing flights. Owners of white-painted WBC hives will often have the chance to observe bees sunning themselves on the side of the hive reflecting the sunlight. Sometimes the bees leave quite a mess on the hive with their droppings. This can also be a problem with bees kept in the garden when the neighbours have washing on the line: white sheets soiled by bee droppings tend to strain neighbourly relations. Fortunately, they are usually unaware of the real culprits which have ruined their washday. In such instances it is best to adopt a low profile and let the birds take the blame.

Balling of the queen

It is unwise to withdraw combs from the hives too early in the year. To do so can result in a phenomenon known as 'balling' the queen. I have never observed this as I make it a policy to leave the bees alone as much as possible in the spring.

An insect that has survived for several million years without human assistance can be safely left to its own devices, given good health and plenty of food.

Balling may occur if the frames are withdrawn or even parted, and appears to be the result of the bees clustering tightly around the queen possibly prompted by some protective urge. Such overzealous protection can badly maul or even kill the queen. Anyone who has played rugby at school knows what it is like to be trampled by the rest of the players in the scrum and can therefore imagine what it might be like for a queen to be on the receiving end of several hundred or even thousand other bees intent on hiding and 'protecting' her. So be a little more patient and leave a proper examination of the hive until at least the middle of April.

Early examination

A thorough examination of each hive may be made from the middle of April onwards. Choose a warm sunny day without a chill breeze blowing because cold can cause the death of unsealed brood. If, because of limited time available, the examination has to be made when the weather is not too fine, do not keep the combs away from the rest of the brood nest for more than half a minute. This is more than ample time and, with practice, ten to fifteen seconds will suffice for the examination of a frame during this comparatively quiet time of the year.

This early examination will reveal whether the queen is present and laying, and demonstrate the size of the brood nest and the extent of the food stores. Everything may be proceeding in a healthy manner or the reverse may be the case. The queen could be missing, the brood nest almost non-existent or the whole colony starving. If a hive has been damp, the outer frames may well be mouldy, and I always make a point of replacing them with fresh wax. Box hives are much more susceptible to damp than WBC hives, assuming they are in good condition. Fortunately, the single-walled hive dries out quickly and the problem soon disappears.

If there is no sign of laying and the queen cannot be located, then you can be fairly sure that she is missing. It is really too early in the year to attempt uniting bees and weak colonies at this time are usually not worth bothering with, but if you only have one or two colonies it is always a difficult decision to destroy them.

When the brood nest area is small, say no more than 4–5in (10–13cm) maximum diameter in the middle frames, although the food stores are adequate, the colony may pick up and thrive, but most likely the queen is a poor layer which may either be hereditary or attributable to her advancing age. Keep an eye on such a colony and artificially requeen or remove that particular queen a little later in the year. From the end of May the bees will be quite capable of raising a replacement queen and, indeed, by that time queen cells may well be already formed.

Starvation is much simpler to remedy, provided that the bees are not already dead! Use a rapid feeder and feed warm syrup during the evening to avoid robbing which may occur with a weakened colony. Such action is often successful but if the colony is excessively weakened the bees will appear listless and even though the food may be taken down there will be insufficient numbers to rebuild the organisation of the colony.

Looking on the brighter side, the bees may well be healthy and thriving and reflecting the care, attention and benefit of adequate winter feeding. If the bees only cover three or four frames, it is worth while feeding them to stimulate growth. If the colony is seven frames strong, they may be supered if required. Keep an eye on them though: late April and early May can be treacherous times of year as far as the weather is concerned, and it is easy to be caught out with a rapidly expanding stock suddenly running short of food. It is a simple equation; the larger the stock the greater the demand on food supplies, both nectar and pollen. It is rather like trying to build a bigger and bigger bonfire that is rapidly burning. The demand for more fuel becomes progressively greater as the size of the fire increases. Do not be surprised if some years you have to feed bees in June to avoid starvation.

Stimulative feeding

Apart from heavy feeding during the autumn and winter and other measures to counteract imminent starvation, all feeding of bees is stimulative. A nucleus colony fed with sugar syrup to encourage expansion is stimulated to growth by supplying readily available food, in effect simulating a nectar flow. Stimulative feeding is only successful if pollen is available and being gathered because this is needed for the growing brood. The term is most often applied to the feeding of colonies of bees that have survived the winter and need more than just spring sunshine to build them back up to peak condition.

Some beekeepers actively practise stimulative spring feeding, while others feel that as long as the bees are not actually starving they are best left to look after themselves. There is no doubt that one can fall foul of the weather by feeding too early; then the colony expands although there may not be any blossom available to work. In such cases a large, strong colony of bees will need further feeding to avoid starvation. It is possible to find bees on the verge of starvation at any time during the summer if the weather has been unkind for some time.

So, it is a gamble. You may apply stimulative feeding early in April when the plum and damson blossom is starting to show. It could be followed by favourable weather conditions during which the bees will be able to work and accumulate nectar and pollen, always assuming that you have located them in a suitable area. It could be a magnificent start to a bumper crop, or it might simply be the prelude to a further round of feeding. If the plum and apple blossom were ruined by late frost—not at all an uncommon occurrence—and this were followed by a wet, miserable May, one would be left with considerable shortage of nectar. Large colonies of bees ready to gather stores but unable to do so, and with only a minimum of stores in the hive, will face starvation unless more food is supplied.

I tend to apply a middle-of-the-road technique which is easy for the amateur to do with only a few stocks of bees. Each colony is examined to determine its individual requirements. It may be small, the brood only spread over three frames, but with plenty of stores. Quite often in such cases the frames adjacent to the brood area are empty but frames further out still hold plenty of food. Simply swap the frames to place the food close to the brood area. This will not restrict a laying queen as the stores will be consumed quite rapidly.

A colony that is small and lacking in stores requires a little more attention. In such cases I would apply stimulative feeding, providing a sugar syrup mix of equal volumes of sugar and water. A contact feeder is best used for this, preferably one with only a few holes which cannot be emptied quickly. This satisfies, to some extent, the replacement of diminished stores and also encourages the colony to increase by artificially reproducing the conditions of a nectar flow. Do not overdo this feeding and finish off with a thicker sugar syrup mix which will provide stores.

Fruit pollination

Bees are amongst the best pollinating insects and adequate pollination is essential for most kinds of fruit trees. Whilst wild solitary bees and bumble bees are useful pollinators, they are seldom present in large enough numbers to be really effective. A hive of honey bees on the other hand may contain many flying bees and has the benefit of being a portable unit. It can be moved into an orchard when pre-blossom insecticidal spraying is finished and removed before post-blossom spraying.

At all cost avoid exposing bees to open blossom that has been sprayed. Some (few, thank goodness) fruit growers are rather indiscriminate with their spraying. I have seen fruit orchards sprayed with blossom already open and hives of bees in the immediate neighbourhood. The result is tragic. Piles of dead bees spilling out of the hive entrance, and the possibility of contaminated stores; it is a heartbreaking sight. What makes it worse is that it is not a freak of nature but the tangible results of another man's selfishness or stupidity.

If you are asked to move bees into fruit orchards for pollination then check with the grower, and his neighbours if necessary, whether there will be any spraying and if there is any doubt avoid the location like the plague. If there is no danger of spraying, then move your bees if you are so inclined: it makes a useful start to the year. Remember,

There is considerable activity in a strong colony of bees during the spring and summer. Even with the entrance block removed, there are masses of bees around the front of the hive. Note the dead bees; this hive was moved some distance without adequate ventilation *(author)*

though, to benefit both beekeeper and grower the colonies should be healthy and cover at least seven British Standard frames. Also, for effective pollination, you need one strong colony per acre of mature trees. It is wise to move the bees to an orchard site after the first flowers have opened. If moved earlier, the bees may have become conditioned to visiting other sources of forage in the area which they will often continue to use even after the fruit trees are in flower.

Do not be disappointed if you show little profit from fruit blossom. There will normally be an increase in the stores but what may be gained in a few days of sunshine can be rapidly consumed in inclement weather. This type of early blossom is a useful natural stimulus to growth but is not to be relied upon. Most years, the gathering of nectar and pollen from fruit orchards is marred by poor weather.

There is always, however, the exceptional year, and 1980 was such a year in our district. The main apiary was located in an orchard and several supered hives contained 5–10lb (2.25–5.0kg) of honey at the termination of the apple blossom. For the rest of the year we were plagued with swarming, even from colonies normally disinclined to swarm. It was possibly a coincidence, but without a doubt it was the best possible start to a year that we had experienced and the worst for swarming!

Polythene tunnel pollination

Horticulturalists are increasingly using polythene tunnels as a sort of portable cheap substitute greenhouse. Most of these tunnels, which vary from 8 to 12ft (2.5 to 3.5m) across inside and can be of any length, are used for growing radish, lettuce and summer greenstuffs. Sometimes they will be used for strawberries which require pollination for maximum cropping. Bees do not fare well in such an environment and colonies become rapidly weakened. Do not use your bees for pollinating polythene tunnels or greenhouses without adequate recompense. In other words, charge double

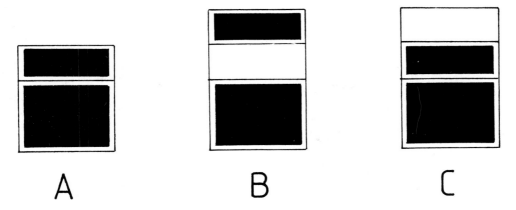

A   B   C

Fig 14 Two methods of adding supers to an already supered hive (A). (B) shows the new super placed under the full super, and (C) shows the new super simply placed on top of the existing full, or part full, super

what you may be offered for orchard pollination which, after all, is of some benefit to the bees. I would estimate that two weeks is sufficient exposure for any reasonable colonies, and at the end of this time they should be removed and replaced with a fresh colony.

Supering

It is difficult to define a date when supers should be added to a full overwintered colony. For the amateur there is a lot to be said for treating each colony on its individual merits, checking that the bees cover at least seven British Standard frames and that they are well stocked with stores before supering. Some beekeepers will tell you to add supers in late March or perhaps early April to give the bees plenty of room and let them get used to the supers. There is a logic to both viewpoints, and it is very much a matter of individual choice or circumstances. One or two colonies in the garden can receive almost daily attention, but thirty or forty hives scattered around the locality will require a different approach.

Friendly debate will always rage amongst any gathering of beekeepers on the subject of adding supers to already supered colonies. You will soon learn that there are two schools of thought amongst the amateur fraternity. One is that it is better to add new empty or undrawn supers immediately above the brood chamber and replace any full supers on the top. The other is that it is only necessary to add the unfilled or undrawn supers onto any existing full supers. There are slight variations to these two basic lines of thought. For instance, some beekeepers would always place a super of undrawn foundation below full supers but add unfilled but drawn out supers above full supers.

By sandwiching unfilled supers between a full super and brood box, the bees do have to travel over the unfamiliar frames and therefore soon impart the colony scent in them. It can be useful to add supers, consisting of frames of undrawn foundation, in this manner. Supers made up of frames of drawn out foundation can be added in either manner, although placing an unfilled super next to the brood chamber does relieve some of the crowding effect at the height of the summer season. Either way we are at odds with nature because bees in the wild expand outwards and downwards, not upwards.

Moving bees

During the active part of the beekeeping season, it often becomes necessary to move hives from one location to another or within the garden or apiary. Bees navigate to and from the hive and this can be observed by moving a hive or even turning it so that the entrance is in a different position. If a hive is moved a few feet in the middle of the day, the flying bees will automatically return to its former location and, not finding it, they will mill around in circles over the position previously occupied by the hive.

Moving the entrance by turning a hive through 180° is almost as confusing to them. Returning bees will cluster at the rear of the hive because this face of the hive is where the entrance used to be. Eventually one or two bees will venture around the hive until they find the entrance and they relay this information to the others by 'fanning' by the entrance. This inability of bees to recognize a hive can be made use of and does not cause any problems either when moving a few feet or several miles, as long as two basic rules are observed.

First, in the garden or apiary hives should not be moved more than 2ft (60cm) at any one time. If it is necessary or desirable to move the hive more than a couple of feet, and this is often the case, then the operation should be conducted in stages, ie at 2ft (60cm) per day. When moving hives over these short distances, do not place a new hive in a position just vacated by a hive. If you do, the flying bees from the first hive will return to the hive standing in its place. Bees carrying food will be accepted into the other hive but this means that the first colony will lose the majority of its flying bees. Sometimes, of course, it is useful to create this situation but it should be avoided when the intention is just to move colonies.

Secondly, when moving full hives from one area to another, ensure that the minimum distance between the new and old sites is 2 miles (3km) and that there is a plentiful supply of nectar and pollen on or very near the new site. When moving sites, it is very much a case of the further the better. If a hive is only moved a quarter of a mile away (400m), the flying bees will return to their old site as they regularly cover such distances. If the food supply being worked is in the direction of the proposed move then add on this distance to the minimum, otherwise the bees may relocate this source of food and, if it was between the two sites, some bees could drift back to the original site even though there would be no hive to receive them.

Move hives after dusk when all the flying bees have returned. In the apiary or garden it avoids having confused bees milling around which happens if hives are moved during the day. If you use the type of hive which has an extended floor or flight board, place a handful of grass across the entrance after moving. The bees are slowed down in their efforts to leave the next morning and it seems to give them time to adjust to the new location.

Transport hives from one location to another after dark whenever possible. As the final preparations for moving, ie sealing the hives, are usually done at dusk it makes sense to complete the move as quickly as possible. Also, moving during the night excites the bees less and is cooler which is also beneficial.

Preparing to move

Moving in the apiary does not require any special preparation, but moving bees several miles requires a little more attention. The best type of hive for moving around is the box hive, and even the most ardent fan of the double-walled hive would probably not disagree. After all, the box hive was designed partly with ease of transportation in mind, unlike the WBC-type which is essentially a static hive. The WBC, in particular, has a nasty habit of shaking apart during transit, and the entrance shutters are rarely beeproof. These hives, with their legs, lifts, roofs and complement of brood chamber and supers, are clumsy to move and take up a disproportionate amount of room.

Transporting a WBC hive in a closed motor car can be particularly eventful. For a variety of reasons, myself and a colleague were engaged in moving two such hives of bees early one morning. The destination was several miles away and, due to unforeseen delays, it became clear that we would not reach it until about 10 o'clock later that morning. The route passed through a small town and at about 9.30am the High Street shoppers were treated to the spectacle of a small estate car travelling west, windows black with insects and containing two persons wearing strange hats with veils. Needless to say, one of the hives had come apart after braking suddenly to miss a farm dog. The local constabulary did not attempt to stop us and investigate! That was not my last experience of transporting WBC hives, but since then I have stapled all the internal

This double-supered hive has been taped together ready for moving. The entrance block will be closed in the late evening, and the hive stood inside the roof for transport. If the hive were to be moved more than 10 or 12 miles, a mesh screen would replace the crownboard to aid ventilation and allow water to be fed to the bees if necessary *(author)*

fittings together, lashed them to the floor and carried the hives in a trailer.

In comparison, the box hive is a delight to transport. We use National hives in the apiary for any colonies that are likely to be moved, and the sequence of operations is as follows:

1. *Late afternoon*: Remove the roof and replace the crownboard with a travelling screen. Screw or staple in place. Staple or tie the sections of the hive together. Replace the roof (at this point I like to loosen the entrance block or remove it completely ready for refitting later).

2. *Dusk*: Remove the entrance block and replace with a closed side. Often when the entrance block is removed a few bees will re-emerge which is why I like to remove the block earlier. It saves time at the end of the day. Remove the roof, invert it and place the hive inside. With a properly fitted roof the entrance block will not come loose. Load on to the vehicle and move to the new site. If the journey takes several hours in a closed vehicle or on a warm, muggy night, sprinkle water on to the travelling screen. At the destination, place the hive or hives on stands, replace the roof and remember to change the entrance block to the open position.

3. *Next day*: Remove the travelling screen and replace with a crownboard; otherwise the holes in the screen will be blocked with propolis. If it is not practicable to go back to the site for a few days then it only means cleaning the travelling screen later or changing it for a crownboard by torchlight on the same night as it is moved.

One hint for the do-it-yourself beekeeper who intends to make his own camping trailer for transporting hives. Make it of a suitable size to take four box hives or more, and remember to calculate the size using the roof measurements because these will be sitting inside the trailer.

Honey or nectar flow

While you struggle with the rudiments of this ancient craft, established beekeepers will talk airily about the honey flow that is about to commence, is in full swing or has just finished. You may be advised to carry out certain operations only during a good honey flow. All the time you are absorbing this additional information you will be trying to deduce what on earth this phenomenon called honey flow is!

There is nothing really mysterious about it. A honey flow is the tangible result of a combination of climatic conditions and plant life that cause the flower-bearing plants to produce nectar. Really we should be talking about a nectar flow rather than honey flow, but we use the latter term as the results are seen in the hive as stored honey.

Even to the beginner, it soon becomes obvious that some flowering plants are more attractive to bees than others. Should the flowering of these plants coincide with the right weather, then a nectar or honey flow is on the way. The observant beekeeper soon becomes adept at spotting plants in his or her neighbourhood which are capable of producing large amounts of nectar and are particularly attractive to bees. There are other factors, of course. Different species of flowers secrete not only different amounts but different concentrations of nectar. Some types of nectar are more attractive to bees than others and the sugar concentration may be a deciding factor. The bee, ever alert to the best sources, will rarely waste time over a handful of hedgerow flowers producing nectar of, say, 25 per cent sugar when there is an adjacent field of oil seed rape producing a sea of nectar at perhaps 75 per cent sugar.

Large areas devoted to the growing of one particular crop, such as oil seed rape or blackberries, can cause a few surprises. One would think that, given good weather, the bees would be more than content faced with a massive harvest right in front of the hive. On a warm summer's day such an apiary is literally abuzz with what the poets would undoubtedly describe as a contented hum.

It is often far from the truth when one opens the hive, as the bees can be a little more than just irritable. The trouble seems to be that when feeding on a common crop some individual hive scent is lost and returning bees drift to other hives. This drifting often occurs but, when a single crop is involved, it may reach a quite unacceptable level. The bees as a consequence adopt a very defensive attitude which can come as something of a surprise when the same colony has been easy to handle the rest of the year.

The following list contains most of the common honey-producing plants of greatest benefit to the beekeeper. Charlock is included although it is not now very widespread where weed killers are in common use:

Blackberry
Charlock
Clover (white)
Dandelion
Field bean
Fruit trees (apple, plum, pear etc)
Heather (bell heather and ling)
Horse chestnut
Lime
Lucerne
Mustard
Oil seed rape
Soft fruit (raspberry, blackcurrant, gooseberry etc)
Sycamore
Willow
Willowherb

7 Summer

Summer is the peak of the beekeeper's season. The weather should be warm and sunny and the bees working well bringing in pollen and nectar. Colonies expand rapidly under ideal conditions and it always gives a great deal of pleasure to see a small nucleus grow in a few weeks to a full supered colony. Of course, it is not quite that easy. Summer in Britain is quite unpredictable and it is rare that any two consecutive seasons are the same. Sometimes the climatic conditions change weekly or even daily, and the observant beekeeper will find that the vagaries of the weather take on a new and more important meaning.

For instance, a warm sunny day may mean that the bees are working hard foraging in the fields and hedgerows, but it depends upon the exact time of year and the availability of suitable crops within the area. In many places the month of June is noted for low yields of honey to the extent that it is often referred to as the June 'gap'. A few dull wet days in the spring or autumn make little difference, but in the height of the summer that same change in the weather may encourage preparation for swarming.

An abundance of hot sunny weather does not necessarily mean a good honey harvest. In England the very hot summer of 1976 was generally poor for the production of honey. It was not a natural climatic condition for Britain. While we sunbathed, the plant life struggled, blossoming into flower and then rapidly going to seed. The flowers formed were unnaturally small although often greater in number like some of the sunflowers I saw sporting seven or eight heads instead of the usual two or three. To the beekeeper it was of little use: it was just too hot. The flowers which formed rapidly shrivelled and wilted, producing little nectar and a reduced amount of pollen.

If we are lucky we may have an average summer. Perhaps we can form a few nuclei and still get a couple of supers of honey from the parent colonies. There is always the exceptional year when one or two hives may be loaded with up to 100lb (45kg) of honey but, taking the good years with the bad, 30lb (13kg) of honey per hive is a reasonable average.

Swarm control

It is probably the dream of every beekeeper to develop or discover a strain of bees that is gentle to handle, hardworking and non-swarming. In practice, while the first two qualities could perhaps be developed, a non-swarming variety of honey bee would be a most unlikely occurrence. The instinct to swarm or prepare to swarm is a natural part of the annual cycle of events within the hive. It is the honey bee's method of expansion, using the principle of divide and multiply. That is not to say, however, that every colony will swarm or attempt to swarm each year. Some bees are much less inclined to swarm than others and, if these are also useful workers, it is worth while considering breeding from them. However, we have to accept that, to varying degrees, our bees will attempt to swarm.

While swarming is a nuisance to the beekeeper, it is not harmful to the bees and should not be regarded as some sort of lemming-like activity brought about by mismanagement. What we try to achieve is a balance between swarm control and the production of honey for it is easy to become overzealous in controlling the swarming aspect to the detriment of honey production. At the same time, we do not want to be plagued with swarms as this will ruin the chances of a good honey harvest since, as well as temporarily reducing the strength of the colony, each departing swarm will take some

A swarm of bees on the move—perhaps some 20,000 insects. This is comparatively harmless, but most people are extremely wary of swarms (*author*)

of the stored honey with it for supplies.

I remember vividly my first colony which swarmed. By the first week in June it was sporting an almost full super of honey and I was looking forward to a prosperous season. I then went on holiday for a fortnight and when I came back they had swarmed, the departing bees having ripped open the centre frames of the super and taken possibly a third of the contents. This early introduction to swarming was useful as it made me aware of the need to exercise some sort of control. Before that, I had simply cut out any queen cells that I happened to find and hoped that nothing more was required. It is necessary to remove queen cells every seven to ten days but such action will not stop the bees swarming if they are determined to do so.

If queen cells are left too long before cutting out and the resident queen still leaves with a swarm, the result may be a queenless colony. If there are no unhatched eggs or larvae less than thirty-six hours old, the bees cannot produce a satisfactory replacement queen. Stunted queens are sometimes produced or even laying worker bees capable only of producing drones. Most often the colony will become queenless and be in danger of dying out.

If you are going to practise cutting out queen cells, then do so as early as possible in their development and this means checking the colonies every seven to ten days and not leaving it any longer. Sometimes this means that examination will be conducted in less than ideal climatic conditions when the bees may be inclined to be spiteful, but that cannot be helped. Throughout the course of the year any number of operations may have to be carried out under somewhat inclement conditions.

Removing queen cells is not foolproof and is indeed only part of the answer to minimising the possibility of swarming. One or two other measures can be taken. First, make sure the hive is not overcrowded by giving the bees plenty of room for expansion. Ensure that there is room available in the supers and, if possible, introduce some undrawn foundation to keep them busy. Remember, though, that while the bees are producing wax and building comb they are not storing honey so a little sacrifice is being made in this direction. Secondly, with a full-sized colony, it may be necessary to open the entrance to the maximum to assist not only the passage of the bees but the ventilation of the hive. Overheating, as well as overcrowding, is conducive to swarming. A young queen in a colony is also a noted aid to lessening the swarming urge, but that same queen needs to be in her second year to be at her best and then there is more chance of a swarm than with a first-year queen.

So, do all that you can to provide conditions that will lessen the chances of swarming: remove queen cells, provide plenty of room, good ventilation and a young queen. Quite often this attention will be adequate, although there are always those strains of bees which will swarm regardless of almost any type of control, and in some years swarming is much more of a problem than in others. Whether this is due purely to climatic conditions is not known, but it can affect apiaries spread over quite large areas.

A reasonable size swarm of bees *(Mr and Mrs R. A. Lawrence, West Midlands Safari Park)*

If you decide to practise a minimum of control, then you must also accept that the bees will swarm from time to time which, if you hive them, means an increase in the total number of colonies. Perhaps you do not wish to suffer an increase. Then dispose of the swarm either as it is or on frames or unite the swarm either with the parent colony or another unrelated colony. This is normally done in early August after the main swarming season which is generally late May to mid-July.

I know now which colonies of bees in our apiary are inclined to swarming and those that can be 'trusted' not to swarm for two or three years even with the minimum of control. Obviously, it is the latter strain which we breed from whenever possible. During the third-year cycle of such a colony, when swarming becomes more likely, we produce a nucleus early in the season, usually of three frames and containing the existing queen, which in effect creates an artificial swarm. The parent colony has foundation added to make up the full complement of frames and is allowed to requeen by raising queen cells. To lessen the chances of a maiden

Collecting a swarm from a post can be comparatively easy, as this sequence shows *(Mr and Mrs R. A. Lawrence, West Midlands Safari Park)*
(above) The white sheet at the ready, and the skep held under the swarm

(below) The next move is to dislodge the swarm, either by giving the post a hefty clout, or by brushing the swarm with a large soft-bristled brush, or by hand

(above) The swarm safely in the skep, turned over on to the waiting sheet

swarm, ie a swarm headed by a newly hatched virgin queen, production of the queen cells is closely monitored and only one allowed to remain for hatching. Once the new queen is laying, the colony soon resumes full production. The nucleus colony with the third-year queen will grow quite slowly at first as all the flying bees were left with the parent colony. It is therefore essential to feed such a nucleus in the absence of foraging bees.

In the autumn the decision has to be made whether to unite the nucleus with either the parent colony or another colony, or feed it for overwintering. I have practised both methods and in the following year have either culled the old queen, now in her fourth year, and allowed the bees to raise another or, if she has proved to be particularly prolific, left her to carry on until later in the season.

Attention to the bees' needs, allied to the simple methods of swarm control which I have outlined, are far from perfect but are adequate for the beginner who is trying to learn a lot in a short time. There are other more comprehensive forms of swarm control which the reader may wish to acquaint him or herself with but such methods have been documented elsewhere and are outside the scope of this book.

Collecting swarms

Bees seem to have a well-developed instinct to swarm at the most inconvenient times and assemble in the most awkward places! They do not have such an instinct, of course, it just seems that way at the time. You may be at work, in the bath or dressed up to go out for the evening when the phone rings and a breathless voice informs you of a 'giant swarm' that has issued forth from your apiary, some other apiary or a hollow tree!

I soon learned to ask for exact details of size and location either for my own benefit or to pass the information on to an associate. To the non-beekeeper swarms always appear to be 'giant'. I have been called out to collect swarms which have ranged in size from little more than a pitiful fistful to those resembling the size and shape of a rugby football. Except when the information has been passed on by another beekeeper, I have never had any of them described to me as 'small'. It can, in fact, be difficult to judge the size of a swarm until one is a little more experienced in the matter. A swarm hanging in a leafy bush or around the forked branches of a small tree always looks larger than it actually is. Take away the foliage and branches and it will appear to shrink quite considerably.

Let us assume that there is a swarm ready for collection and conveniently situated on the end of a branch perhaps on a small bush. Apart from the normal protective clothing, the only equipment needed is a skep or box (a cardboard box will do if sound), a large cloth or piece of sheeting and a jar of honey (not imported honey because of the possibility of disease). What may be regarded as the classic approach to taking a swarm is carried out in the following series of simple operations:

1. Place the sheet on the ground either directly under the swarm or, if this is not convenient, a little to one side.

2. Spread honey from the jar around the inside of the skep. This can be done with a twig or other handy piece of wood, but I usually have the hive tool in my overall pocket and use it as a spreader. Do not overdo this application of honey; it only needs a thin smear to make the skep attractive to the bees. Many old-time beekeepers had their own favourite recipes: dusting the inside with a handful of flowers was often used and works quite well. I favour honey because it is consistently attractive. For several years I had a large jar of cappings which was used for this purpose; mixed with the residual honey it made a soft white paste and proved very useful. On other occasions I have resorted to spraying sugar syrup inside the skep or box. Theoretically, no such attraction is needed but in practice it can be very useful. One word of warning: applying honey or cappings is best performed while standing some distance away otherwise a surprising number of bees will find their way into the jar.

3. Hold the inverted skep right up under the swarm, then give the branch a hefty blow to dislodge the swarm.

4. Quickly place the skep in the middle of the sheet, turning the skep mouth down as it is placed. The bulk of the bees will fall down onto the sheet but as long as the mouth of the skep still covers them it does not matter. I usually drop the skep onto one edge of the mouth then it only needs turning through 90°.

5. Prop open one side of the skep with a stone or twig, about $\frac{1}{2}$in (13mm) or so is sufficient. This allows the flying bees and those remaining on the branch to join the rest of the swarm. What we have done, in effect, is to create temporary accommodation but as far as the bees are concerned it is now their home and, if left long enough, they will start to build natural comb.

6. Leave the skep until the evening. As long as the queen is in the skep, the bees will almost certainly stay where they are.

Put thus, it all sounds very easy, and under ideal conditions as described it is. I regard swarm collection as the party piece of beekeeping, a useful public relations exercise. Swarm collection attracts a good audience; most observers are really fascinated and almost refuse to believe that anyone with a little knowledge can do the same thing. Perhaps they subconsciously equate white overalls and veil with white-hooded robes and other mysterious mumbo-jumbo!

The situation can be a little more difficult when a swarm lands in the middle of a hedge, on a post or under a branch. This difficulty can be compounded if the swarm has been in place for two or three days when the bees may be hungry, especially if the weather has been cold and wet since they left the hive.

When on general swarm collection, in addition to the basic equipment, I always carry a pair of garden secateurs, a large soft-bristled paint brush and the smoker. If the swarm is in a hedge, the secateurs are often invaluable to cut away some of the foliage to get a clear view or to enable the skep to be pushed into place, mouth down and directly above the bees. All that is needed then is a little smoke and a lot of patience to drive the bees up into the skep. This little job can sometimes last a couple of hours.

A swarm lodged on a post can be dealt with in a similar manner, but one hanging under the branch of a tree calls for a slightly different technique. In such instances one has to be a little more brutal. Holding the skep with one arm as close to the swarm as possible enables one quickly to brush the swarm into it. Quite often this is successful and only a comparatively small number of bees take flight, the majority falling into the skep with a satisfying thud. However, if the queen has been missed, or sometimes for no apparent reason, the swarm will leave the skep almost as quickly as it entered. Then you learn the meaning of perseverance!

A season of swarm collecting is very useful experience, and I have always regarded it as one of the most satisfying aspects of my beekeeping 'apprenticeship'; I soon learned a few shortcuts, usually born of necessity.

Swarms close to home did not normally constitute a problem. They could be collected in a skep and picked up later that evening. Swarms several miles from home were dealt with in one of two ways. The favourite method was to take a nucleus hive fitted out with two frames of drawn out foundation. If the swarm was in an easily accessible place, it could be shaken straight into the open hive with the frames, previously sprayed with sugar syrup, placed one on either side of the brood area. With the swarm in the hive, the frames were pushed together and the crownboard and roof fitted. If the immediate area was vandalproof and free from large animals, the hive could be left and retrieved up to a couple of days later. There is always the chance that the bees will build natural comb to one side of the hanging frames but that can be dealt with at leisure. This method of swarm collection is also quite suitable for use in the apiary but it can only deal with easily accessible swarms.

Almost inaccessible swarms had to be collected in a skep or box and if, for a variety of reasons, they had to be moved almost immediately I used to adopt an emergency measure which had a limited success rate. The swarm would be dropped or induced into the skep and quickly inverted onto the waiting sheet which would then be wrapped around the skep and the loose ends tied together to make a reasonably beeproof package. The skep was then carefully reinverted and the sheet stretched over its mouth sprayed with water to aid cooling by evaporation as well as giving a parcel of hot excitable bees something to drink. A speedy return home was then essential, and on arrival the skep, sheet and contents were hung in a cool dark place and usually sprayed with water once again. The swarm would be hived in the late evening, in fact as late as possible, since after this rather rough treatment they would, when let free, take umbrage and depart from the site of their intended home as quickly as possible.

This rather crude method of bee handling does entail some losses. Bees are left behind where the swarm was lodged, but if they are within a short distance of their original hive, then they will eventually drift back. If, however, the swarm was in place for a couple of days, this may not be the case. Other losses are incurred in transit when the bees become excited, and in the worst case one can finish up with a soggy mass of crawling bees. To minimise the chances of this happening I use the largest available skep or box to enable the

"COLD"

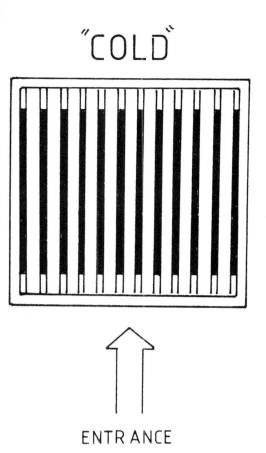

↑

ENTRANCE

"WARM"

↑

ENTRANCE

Fig 15 The so-called 'cold' and 'warm' methods of placing frames in a hive relative to the entrance. The 'cold' method is the most commonly used and often the most convenient

bees to spread out as much as possible. This emergency measure is not one to be practised too often but can be useful at times.

Do remember that if you lose the queen or inadvertently kill or seriously damage her, the swarm is virtually useless. If the queen is not dropped into the skep with the rest of the bees, they will quickly leave it and reassemble with her wherever she is.

Hiving swarms

Hiving a swarm when it has been collected in a skep or a box is a most satisfying job. It is best carried out in the evening of the same day in which the swarm was captured. A simple routine should be followed:

1. Select an empty hive, making sure it is clean and in good repair.

2. Add to the hive four or five frames of drawn or undrawn foundation smeared with a little honey or sprayed with sugar syrup. Fit a division board and blanked-off crown-board.

3. Place a piece of wood (another crown-board will often suffice) against the front of the hive leading up to the entrance. With WBC hives open the entrance slides to the full extent and with box hives remove the entrance block completely.

4. Pick up the skep by drawing up the loose ends of the sheet on which it is sitting and place it mouth downwards on the sloping board.

5. Roll the sheet down to expose the skep making sure no fold of sheeting obstructs the entrance to the hive.

6. Gently lift up the skep then give it a good hefty downward shake to deposit the

bees on the sheet laid across the sloping board. The bees' instinct will make them crawl upwards, and when the first few scouts find the entrance they will carry out a quick examination inside and then start fanning to attract the mass of bees milling about in an undecided fashion. As the bees receive the message, they will turn towards the hive entrance and move in that direction. Once the queen has entered the hive, there is no holding them back, and it is a wonderful sight: row upon row of bees streaming into the hive, others fanning at the entrance to encourage the others. Every time I hive a swarm in this manner and watch them, at first a little confused and undecided and then suddenly committed and pouring into the hive, it gives me a distinct glow of pleasure.

7. Close down the entrance of the hive to a minimum and remove the sheet and board from the entrance.

8. Place a feeder on the hive, preferably one holding at least 4 pints (2 litres), fit the roof and leave for a day before disturbing them again.

Hiving a swarm in the evening makes sense for several reasons. First, there is less chance of robbing or disturbance by other bees. Secondly, it encourages the swarm just ejected from the skep to seek shelter, whereas at an earlier hour in the day the queen may take flight again and the rest of the bees will of course follow. An evening when rain threatens is ideal. I have had the odd swarm which has been reluctant to enter a hive; then the first few spots of rain have fallen and they have almost rushed into the hive to avoid the impending downpour.

Sometimes a swarm will get the wrong side of a poorly fitted division board and build natural comb if left to their own devices. To avoid this, check the hive a day later to make sure the bees are on the frames provided. If not, they will be hanging from the crownboard and I usually remove the division board, part the frames slightly and shake the bees onto them. A little later, if the bees are on the frames, I slide them back together and replace the division board.

This is not normally much of a problem, and with hives where the brood chamber can be turned at 90° to the entrance it should almost never happen. Let me explain: with a full colony I always arrange the frames in the National hives at 90° to the entrance, what old-time beekeepers used to call the 'cold' way. When hiving a swarm on four or five frames, I place the brood box so that the frames are parallel to, or across, the entrance. With a division board fitted at the rear of these frames, access to the empty section of the hive is denied to the incoming swarm. This frame arrangement is known as the 'warm' way, and is very useful when keeping colonies of up to six frames in a full-sized National hive.

Making a nucleus

One of the aspects of keeping bees which the beekeeper soon accepts is the possibility of increase by swarming. While we exercise a certain amount of control to minimise the possibility of swarming, there are times when we desire an increase in the number of colonies, either to enlarge the apiary or to sell a few bees for profit. Breeding bees is often more profitable than honey production, although it is useful to pursue both ends if possible.

A successful nucleus can be made up any time between late May and the end of June. After this, the chances of a small colony expanding sufficiently to go through the winter is rather slim. I have made up nuclei as late as mid-July but they were usually produced by splitting a stock and ensuring the queenless half had a queen cell present. Even this does not guarantee success for if the queen fails to mate successfully she will only be capable of producing drones and this, late in the year, means that the nucleus colony will usually have to be reunited.

If an expansion programme is decided upon, make the necessary moves as early in the breeding season as possible and do not worry about getting honey from a nucleus stock that same season. It is much better to build up a strong stock carefully with plenty of natural stores for wintering rather than stretch them to the limit for a few ounces of honey.

It makes sense to breed from your best bees, ie those colonies which pick up quickly in the spring, are gentle to handle and also

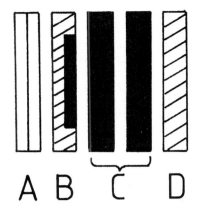

Fig 16 The composition of a nucleus. (A) undrawn foundation or drawn out empty comb; (B) comb containing brood and food; (C) two combs of brood; (D) food

store plenty of surplus honey. If you only have one colony of bees, then no such decision is necessary; it is just a matter of getting on with the job.

First, there is the extra equipment to be brought out of store. The items required to make up and hive one nucleus colony are as follows:

1. Nucleus hive or full-sized hive fitted with a division board.
2. Frames of undrawn foundation, sufficient to replace those full frames taken from the parent colony and two to add to the nucleus colony.
3. Feeder and at least 4 pints (2 litres) of sugar syrup.

In the pursuit of standardisation, I find it useful to use full-sized British National hives for rearing nuclei. I do have one fitted out with a central division board and a pair of half crownboards. It is used with a modified floor which sports an entrance block at either end but with the entrances diagonally offset. Two five-frame colonies can be accommodated side by side as the respective entrances are at opposite ends of the hive. This arrangement has the advantage that the hive roof, brood box and floor, when blocked at one end, can be used with other standard equipment.

An alternative method is to use a full-sized hive with a limited number of frames and a division board sealing off the unoccupied portion of the hive. This is handy if the colony is to be kept, then extra frames can be added as necessary until it reaches full size. Another less obvious advantage is that large feeders can be used, whereas nucleus hives have a limited amount of space for a decent-sized feeder.

Whatever hive is used, the principle is the same: a number of frames are taken from a full colony and installed in another hive allowing the bees to rear a new queen which, it is to be hoped, will subsequently be successfully mated and continue the life of the new colony.

Let us now consider the novice beekeeper with one colony of bees who wishes to make an increase and has at his or her disposal an empty five-frame nucleus hive. It is early in June and a warm sunny day; the bees are flying well and bringing in plenty of pollen. On examination, it is found that the colony has built up until the brood nest reaches across nine frames and there is evidence of a nectar flow in progress by the amount of honey stored around the brood. Some frames are filled with capped brood whilst in others grubs and eggs can be seen. This is an ideal situation, so to work!

1. Place the nucleus hive at the side of the parent hive, the closer the better, touching if possible.
2. Have to hand a number of frames of fresh undrawn foundation. By this time you will probably have decided how many frames are going to be taken from the parent colony. Two will work but is a little risky and five will probably deplete the parent colony more than is desirable, so three or four will be the best number.
3. Remove the roof and crownboard of the nucleus hive.
4. Select from the parent colony one frame with predominantly capped brood, one with grubs and a reasonable quantity of eggs visible and another from the outer edge of the brood nest with brood and plenty of food visible. Place these in the nucleus hive with the frame from the outside of the brood nest placed against a frame of undrawn foundation.
5. Fill the remaining space in the nucleus hive with a comb of food from the outside of

the parent colony and a frame of undrawn foundation to the outside. As each frame is transferred carefully, examine it and make sure the queen is not present. Pay particular attention to the frame containing cells with eggs.

6. Take a further three or four frames and, after examination, shake some of the bees onto the top of the frames in the nucleus hive. This will cause a little excitement for a few seconds and a good many of the bees will return to the parent hive.

7. Replace the crownboard, add the feeder and roof and move the nucleus hive at least 10ft (3m) away from the parent hive.

8. Move the remaining frames in the parent hive together and place two frames of undrawn foundation on either side. Replace the crownboard and roof; this larger colony will not require feeding.

At first, the nucleus colony will seem to weaken quite rapidly as young bees emerge and are not replaced. It will take the bees a little while to realise they no longer have an active laying queen so do not expect a queen cell or cells to appear overnight. If, after three days, there is no sign of any queen cells check again a couple of days later. If the situation is no different, then the new colony must be found a queen or it will perish.

Usually all goes well and a queen cell or cells will be successfully formed. Often the bees will look after themselves, and if several queen cells are formed the first queen to emerge will kill the others still captive in their cells. There is always the chance that she may leave the hive at the head of a swarm, so the cautious beekeeper will destroy all but one queen cell that is formed. Once the young queen is successfully mated, she will start laying and the colony will grow quite rapidly. Do not spare the feeding, though, as plenty of food will do no harm and attention to feeding is always repaid in the long term as the colony grows.

As the colony increases and spreads across to work the undrawn foundation, it will be necessary to give the bees more room, and the only way to achieve this is by transferring them to a larger hive. The new hive should be positioned in the place of the nucleus hive and the frames transferred in the same manner that they were first taken from the

original parent hive. Two more frames of undrawn foundation should be placed on either side of the expanding colony. Add only enough undrawn foundation for the bees to manage at any one time.

If a small colony is subjected to robbing by other larger colonies, close the entrance to the absolute minimum. This is easily done with a double-walled hive with sliding shutters, but box hives fitted with an entrance block may need plugging with grass to reduce the entrance. By the time the grass has withered, the robber bees have usually lost interest and the resident colony will clear out the entrance.

Robber bees do not look different from any other bees, but they can be identified if you see an excited mass of bees hovering around the entrance of a small or weakened colony. Watch for signs of excitement around the entrance and see if some of the milling bees make a dash for the entrance. A legitimate homecoming bee will zoom in straight to the entrance, but a robber will often be part of a mob milling around and waiting for an opportunity, and such a bee will not be laden with nectar or pollen! The robbing frenzy is difficult to describe adequately, but once you have seen it you will understand exactly what I mean. Do not confuse it with a colony preparing to swarm, for then a large number of bees will fly around in front of the hive but activity at the entrance is less obvious.

There is a short cut to raising a nucleus. If the parent colony already contains queen cells then take a suitable frame complete with a queen cell for the nucleus colony and destroy all remaining queen cells. This way, one can be sure that a virgin queen will shortly be present in the nucleus hive and it will save some time before it gets into proper operation.

When a new queen is reared, it is worth while to mark her with the appropriate colour code for the year. This is an international colour code and is used in the following sequence:

Colour	Year ending in
White	1 or 6
Yellow	2 or 7
Red	3 or 8
Green	4 or 9
Blue	5 or 0

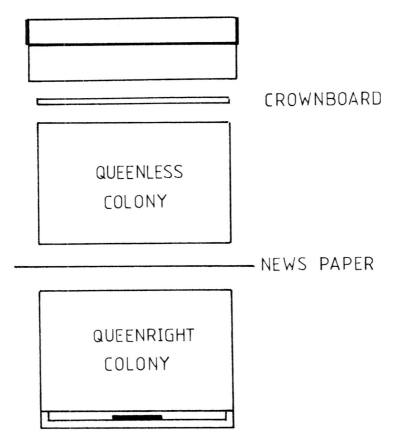

CROWNBOARD

QUEENLESS
COLONY

NEWS PAPER

QUEENRIGHT
COLONY

Fig 17   Uniting two colonies by the newspaper method

From this, we can deduce that the colour code for a queen born in 1985 will be blue or yellow for one born in 1987. Queen-marking kits are available from various suppliers. The marking is applied to the back of the thorax which, when the colour code is white or yellow, has the advantage that she can be more easily located against the dark mass of the other hive inhabitants. From then on, the age of the queen will be known, and if she is replaced by another it soon becomes obvious.

While marking the queen, which can often be done without handling her, you may wish to clip her wings to prevent her leaving or at least going very far in the future at the head of a swarm. Although a queen is a fairly hardy insect, I would not advise a novice to attempt wing clipping until they have seen a practical demonstration by an experienced beekeeper. Even better, if you have faith in the other person let him or her show you how to clip a queen's wings by doing your unclipped queens. That way you will have seen how to do it and completed the exercise at the same time.

Uniting colonies

It is sometimes necessary to unite colonies, usually because one is queenless. It is normally carried out in the autumn as, in the early part of the year, other remedial action can be taken to replace the missing queen.

With full-sized colonies, I prefer the 'newspaper method'. This consists of placing one colony above another with only a sheet of newspaper between them. The bees in the queen-right colony are at the bottom and they continue with their daily toil quite undisturbed. The bees in the top brood box do not have any way out except to eat through the newspaper separating the two colonies. By the time they have eaten their way through, the hive odours have intermingled and they will unite with a minimum of disturbance.

It is not really a practical proposition to place the two colonies together during the daytime as there would be too much disturbance and a lot of flying bees would be away from the queenless colony. It is best performed after dark by torchlight, preparing a suitably sized piece of newspaper beforehand. This can be placed directly over the brood on the receiving hive but it is better if a super is in place. The hive containing the queenless colony is then lifted off its floor and placed onto the newspaper. A few bees will be lost in the process but that is preferable to losing the whole colony. One precaution I take is to free the floor from the hive to be moved sometime earlier in the day. Then, during the actual transfer the brood box can be lifted off with the minimum of disturbance.

If the colony is placed over a supered hive, there will usually be a queen excluder in place. This is all the better because, as the remaining brood hatches out in the queenless colony, they will drift down into the lower brood box and eventually the top brood box will only consist of empty frames and perhaps a little stored honey which can be extracted later. These empty frames, if in good condition, can be packed away and used the next season in place of undrawn foundation.

Of course, it may be very late in the season and then it is simply a case of placing the two brood boxes together with the newspaper between and again with a queen excluder in place if it is wished to empty the upper box eventually. On the other hand, one can unite the two colonies with nothing but the newspaper between them and split them again in the spring to reform two colonies, although they will contain a mixed bag of bees. Do not be surprised, if this is the plan, to examine them in the early spring and find the combined colonies have moved up into the top box leaving the lower brood box almost empty.

When the apiary becomes cluttered with small colonies formed from secondary swarms or nuclei that have not fared well, it is useful to combine them into full-sized colonies. Move the colonies earmarked to be

united close together. If, say, it requires three to make up a full complement of frames, then stand the three side by side. Three small hives or colonies may contain more than sufficient frames but quite often some of the outer frames will be underused and can be removed.

Each colony may contain a laying queen so it will be necessary to cull two of them. When you are not sure of the history of each colony concerned, it is probably best to keep the largest queen which seems to be doing her job reasonably well. Then comes the interesting bit. If the middle hive is a full-sized hive it is a help, but if all three are nucleus hives a full-sized hive should be placed behind them. With all the colonies to be united opened up, spray them liberally with sugar syrup. As each frame is removed, spray both sides until there are a lot of sticky bees wandering around. Place the frames into the full-sized hive and do not be afraid to jumble them up. It is best to put all the large patches of brood together and any scattered brood and food to the outside. Put the crownboard in place quickly before a mass robbing session breaks out, and move the emptied hives out of the way. Any bees still on the floor of these hives should have been shaken onto the tops of the frames before covering them with the crownboard. This operation is best carried out when a good nectar flow is in progress as it reduces the inevitable squabbling which takes place when three different lots of incoming bees meet at the hive entrance.

I find this method of uniting great fun and it works because the sudden abundance of loose food overcomes the bees other instincts and probably masks the individual colony odours. At the same time, the bees are thoroughly disorientated and, after cleaning each other off, the separate colonies are fairly well mixed. Using this method, I have made up a colony from a small swarm established on three frames and added a further six frames of bees, food and brood from six different hives. Two previously drawn out but empty frames were placed at either side of this mixed bag to fill the hive. In this instance it was not necessary to move any hives together. The resident three frames, which were already in a British National hive, were sprayed and each frame which was taken from another hive sprayed before adding it to the hive. One word of warning: speed is essential to avoid robbing, and it is best to close the entrance down until the bees have settled down. To help matters, I complete such operations some distance away from the main apiary whenever possible.

Double brood colonies

Sometimes one plans to produce a double brood colony and at other times it is a matter of circumstance, perhaps a result of uniting two colonies in late autumn. When planned, it is usually to accommodate a particularly active laying queen; this sort of arrangement being much more common with the smaller type of hive. The British National hive with a full complement of eleven British Standard frames may be filled to the limit with brood if the resident queen is better than average, while a ten-frame hive such as the WBC is often too small to contain the laying potential of most decent queens. It may be argued that a brood chamber crammed with brood is not a bad thing because then most of the surplus honey will be stored in the supers, but at the same time a cramped environment is quite likely to trigger off the swarming instinct.

Do not be in a hurry to build double brood colonies. It is very easy to overestimate the laying potential of a queen and a hive which appears to be bursting at the seams in July can look significantly smaller by late August. During the height of the summer there is a lot of activity in a hive and, if the weather is hot, the bees will be fairly well spread out in the hive and this can give a false impression of the true size and strength of a colony. Only from examination of the brood spread across the frames can the true performance of a queen bee be determined, and the larger hives, such as the Langstroth, Modified Dadant and Modified Commercial, usually have sufficient brood area for most reasonably prolific layers.

But, far be it from me to deter you. If you have a WBC, Smith or British National hive and find it crammed with brood right across the hive, then you can consider doubling up the brood area. Be prepared for perhaps a little disappointment, though, as twenty-two British Standard frames of available laying space take a lot of filling. Often, in practice, it is quite beyond the queen bee's capability to make full use of the extra space and one finishes up with a lot of honey stored in the brood frames. It is easy to fall into the trap of thinking that a double brood colony will produce twice as much honey as a single brood colony or the equivalent of two single brood colonies. Of course, this is not so, for only a single queen is present and she is unlikely to have a laying rate equivalent to the combined efforts of two other queens. Without twice as many eggs to develop into the same proportion of worker bees, a double brood colony cannot compete for honey production on equal terms with two single brood colonies.

The effect of a long deep brood area with superimposed brood boxes is a similar layout to comb built in the natural or wild state. Whether this is an advantage I do not know, but the only obvious benefit I found was a lessening of the problems of swarming which may otherwise have been caused by overcrowding. Handling such a large arrangement is more difficult, as is tracking down the queen.

When examining a double brood colony, it is useful to have a spare crownboard to hand. Lift off the roof and place it upside down (if a box hive) and drop any supers diagonally across it with the original crownboard left in place. The top brood box is examined in the normal fashion then lifted off and dropped on top of the supers. This is where a spare crownboard is useful. It is placed on top of the brood box which has been removed from the hive, effectively sealing it off and reducing the amount of flying bees in circulation. The bottom brood box can then be examined, and reassembly is effected in the reverse order, except that I normally lift the top brood box into place before removing the spare crownboard. I confess that I am normally fortunate enough to have a colleague helping and he will remove the spare crownboard just before I drop the supers in place.

There is little doubt in my mind that a supered double brood colony is a daunting prospect for the absolute beginner but a

89

useful educational project in the long apprenticeship of beekeeping.

Requeening

Requeening is performed either to replace an ageing or deficient queen bee or to alter the characteristics of a colony. By requeening it is possible to change an aggressive colony into one that is tolerant and easy to handle. This is possible because a mated queen of a different strain will produce bees which are not related to those in the hive into which she has been introduced. If a virgin queen was introduced it would be a different matter as she would be likely to mate with a drone from that same hive, in which case her offspring could be as savage as those of the queen she replaced. It is difficult to predict what will happen in such circumstances unless one has a considerable knowledge of genetics, so it is advisable to use a mated queen from a known strain.

Introducing a new queen into a colony does have its problems and, unless one is careful or very lucky, she may not survive. Even when extreme caution is exercised things can still go wrong and one can finish up with a temporarily queenless stock. After all, the queen we may be seeking to introduce into a colony is a stranger to that hive and will also carry a different odour.

It is possible to purchase a queen bee of a known strain that should, for her lifetime, produce bees with known characteristics which should be a beneficial influence for future generations of bees. Although the idea of a 'super' queen is very attractive, I think it is wise to err on the side of caution and try ones hand at requeening on a homebred scale before spending several pounds on one vulnerable insect. Of course, if you only have one colony a bought-in queen will be necessary, but before sending off your cheque ask other beekeepers in your area if they have any spare queens. In the latter part of the season you may be lucky.

With more than one hive of bees you can rear your own queen from a favourite colony. There are several ways of doing this, some of which require a considerable amount of manipulation. The easiest method is to raise a small nucleus—two frames will suffice—but take care because such a small colony needs to be monitored closely for signs of robbing or excess weakening. With luck, you may find a frame in the parent colony with a queen cell already attached. If so, use it to form part of the nucleus. Make sure it is a full-sized cell as many part-formed so-called 'play cells' will not be used to form a mature queen cell.

When the nucleus colony is queen-right with a strong laying new young queen, cull the old queen in the colony destined for requeening. Leave this colony for three days, then examine and destroy any queen cells which have been formed. By now the inhabitants of this hive will be quite aware that they are queenless. In their desperation to raise a queen they are much more likely to accept a new arrival than if the old queen had only just been killed.

Now comes the interesting part. The young queen must be caught and transferred to a suitable cage. Most beginners are reluctant to catch the queen as they are afraid of damaging her and, although she is fairly hardy, I can well understand such reluctance. If you are in any way apprehensive, there are a number of proprietary items available to aid the capture and examination of queen bees.

The queen is introduced in a cage to the queenless colony so that those bees can, in effect, become acquainted with the fact that she is indeed a queen and therefore a desirable addition to the hive. Again, a number of proprietary items are available for this purpose but it is possible to make your own queen-introducing cage or adapt some household article. I have used $\frac{5}{8}$in (15mm) diameter plastic hair rollers of the type which are like a perforated tube with one end closed. The elastic straps and ball arrangement used for holding the roller in the hair is dispensed with. With the queen inside, the open end of the roller can be sealed with a piece of newspaper tied across the end or it may be plugged with candy. Whatever is used to close the cage it must be something the bees in the hive can remove— but not too quickly.

The queen-introducing cage is hung down between two frames in the brood box of the queenless colony. The introduction is best

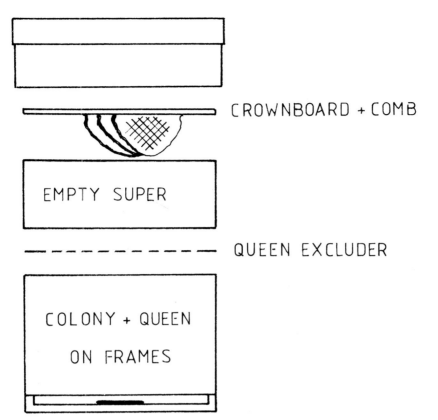

CROWNBOARD + COMB

EMPTY SUPER

QUEEN EXCLUDER

COLONY + QUEEN

ON FRAMES

Fig 18 Making use of natural comb containing brood. As long as the colony has some spare frame space the natural comb will not be extended. Food will be removed to the brood chamber and any brood tended until it hatches. Drones will not be able to pass down through the queen excluder

effected in the quiet of the evening; then leave the colony alone for about a week. If things have gone well, the resident bees will have examined the queen in the cage, released her by eating through the candy or newspaper and accepted her. After a week she should be laying, but if there are no signs of this activity or the queen cannot be found then it is likely that the colony has rejected her and she is dead. A much safer way to requeen with a homebred queen from a nucleus is to unite the two colonies by one of the methods previously described. This is the way we always requeen in the apiary with queens reared on site and it has a very high success rate.

Let us consider now the colony that has become queenless through some mischance or perhaps has an old queen incapable of laying. Sometimes a colony of bees will requeen themselves without swarming. This process is known as 'supersedure' and usually takes place in the autumn; for a while the colony may possess two queens, although the old queen is no longer a threat to the new one. Sooner or later the old queen will disappear either through natural causes or because the bees have discarded her in favour of the new queen. Unfortunately, we cannot rely upon the bees to perform this supersedure, although some colonies are more inclined to do it than others. So, we may be faced with a queenless colony in danger of dying out and no replacement queen to hand. In such cases a frame containing brood and eggs may be transferred from another hive.

Select a frame from a queen-right colony, one with a high percentage of unhatched eggs visible. Look for the queen, and if she is on that frame put her back into the hive. If she proves nervous and elusive and you feel disinclined to pick her up (after all you will not want to risk losing another queen), then place the frame diagonally across those in the hive and smoke the bees down into the brood nest. A nervous queen will soon scuttle away between the frames to hide. After ensuring the queen is safe, any remaining bees can be shaken off the frame with a sharp downward movement. Now, back to the problem. A colony that has been queenless for some time will not have much brood left and the empty cells vacated by the hatched brood are often partly filled with honey. It will not be difficult to find a frame of old underutilised comb in the queenless colony which can be removed and discarded. Fit the replacement frame in or near the middle of the remaining brood nest adjusting the existing frames as required. Check a few days later for queen cells on the new frame. This is a fairly reliable method of encouraging the bees to raise a new queen and is not fraught with the problems that can beset other methods of requeening.

Natural comb

There are a number of occasions when one wishes to make use of natural comb, usually that containing a high proportion of brood. It may be a wild nest cut out of a tree or, more likely, comb built the wrong side of a division board, between incorrectly spaced frames, or in an empty super placed over a hive to contain a feeder or candy to which the bees have gained access. Natural comb built in the hive will be attached to the crownboard and those pieces too small to consider building into a frame may be dealt with in the following manner.

1. Carefully lift up the crownboard to which the comb is attached. Try not to tip the crownboard as this places undue strain on the anchorage point of the comb which may cause it to fall apart.

2. Place the crownboard onto an empty super or brood box, whichever is suitable to accommodate the depth of comb.

3. Fit a queen excluder over the existing brood box and place the super or brood box with crownboard on top. The intention is to trap the queen below the excluder to stop her laying in the natural comb. At the same time, the brood in the natural comb will be attended by nursery bees and young bees will eventually emerge leaving empty comb. This is preferable to just throwing away perfectly good brood. It is necessary to ensure the queen is in the bottom brood box otherwise the bees will expand the natural comb to accommodate her laying. It is useful to get an associate to hold the crownboard and comb over and just touching the frames in the hive, while you smoke the comb. Often, when so treated, the queen, if she is on the natural comb, will run off into the brood box for security.

4. When all the brood has hatched, the comb will often contain honey and still be covered in bees. Lift off the crownboard and super or brood box replacing the queen excluder with a crownboard or clearer board fitted with bee escapes.

5. Replace the crownboard and super or brood box over the clearer board. The bees will then filter back into the original brood box leaving the comb clear of bees. It is useful, if the crownboard or clearer board is the type with a shutter, to let bees pass back through into a super. The natural comb cleared of bees may contain honey which can be distributed to friends in the comb form. Normally, though, it is not capped and the cells are only part filled. If the bees are let back into the super or brood chamber containing this comb they will transfer any honey and pollen to the brood chamber. A couple of days is plenty of time before closing the shutter and letting them filter back into the brood nest via the bee escape. When the operation has been completed, the hive can be reassembled in the conventional layout. If the natural comb was built because of faulty frame spacing, this should be attended to before the natural comb is dealt with as described, otherwise the problem will recur.

When there are several large pieces of natural comb well filled with brood and stores, these can be fitted into empty frames. It is best to clear them of as many bees as possible by smoking the comb and running them onto conventional frames already fitted into a brood chamber. The comb should then be carefully broken away from whatever it is

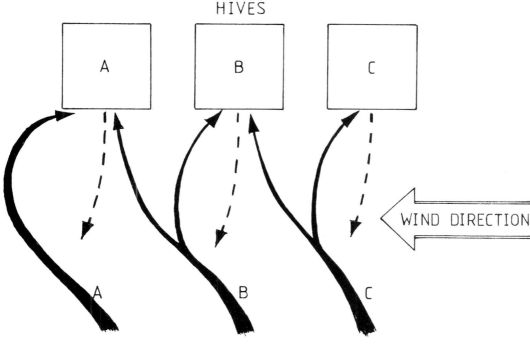

HIVES

WIND DIRECTION

FLIGHT PATH OF RETURNING BEES

Fig 19 The effects of drifting. Hives standing too close together are often subject to a small amount of drifting. A strong crosswind can result in excessive downwind drifting

attached to as a large piece of comb built naturally in a hive may well be fixed to the brood box walls and hive floor. A large sharp kitchen knife is often useful in these circumstances.

I used to try to cut the comb with a large pair of scissors to fit an empty frame. After a few attempts, I found it easier to press the comb into the frame by holding it with one hand against the frame and using thumb and forefinger of the other hand to push it into place. The surplus comb around the edges can then be cut away. Some bees will be lost or embedded in the wax and a certain amount of brood will be squashed, but it is a small sacrifice to pay for saving most of the brood.

When the comb is in the frame and trimmed to fit, I like to tie securely two or three windings of strong thread or nylon fishing line vertically around the frame and comb. This makes sure that it does not fall

out when the whole assembly is put into a brood box. Eventually the bees themselves will fix the comb into the frame and tidy up the tatty edges. It is not advisable, though, to hold an unwired brood frame at 90° to the vertical otherwise the comb may separate and fall out causing a certain amount of excitement!

Drifting: control and use

An ever-present problem in the apiary is that of bees flying to the wrong hive. It does not happen to a large extent but when it does it may spread disease and can instigate robbing or cause the loss of a queen when flying for mating. Drifting, as it is called, is caused mainly by siting the hives incorrectly, perhaps too close together, a fault which can be compounded by also having them in regimented straight lines. It always looks very neat to have them laid out in such a manner, but is not necessarily the most efficient arrangement. A much better layout would be to have the hives in a circle with the

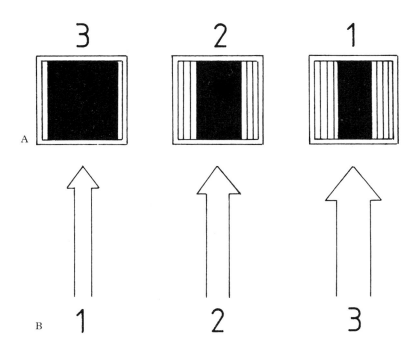

Fig 20 (A) Three hives of differing strengths, and (B) a proportional number of field or flying bees. To build up colony 1, it is swapped with colony 3. Colony 1 then receives a massive influx of flying bees. This kind of hive swapping should only be done when there is a strong nectar flow in progress, and relies on the bee's inability to recognise it's own home hive

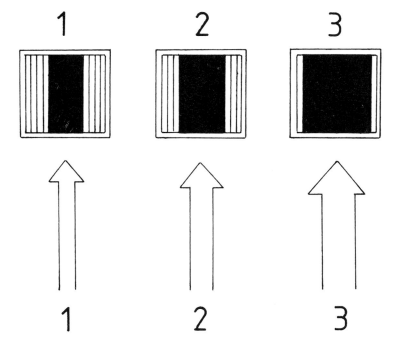

entrances facing outwards, which is perhaps a good arrangement for the beekeeper when working from hive to hive. In practice, except with a fixed number of hives or a well-spaced circle, such a set-up could cause difficulty when changes were planned, eg when raising nuclei. As with most things in life, a compromise is possible and a staggered layout of hives works quite well.

Drifting, or rather the inability of bees to recognise their home, can be useful. A handy trick to aid the growth of a nucleus colony is to leave it in the position formerly occupied by the parent hive, moving that hive several feet away. By doing this the flying bees return to the nucleus hive and bolster its complement of bees quite considerably. The parent colony takes a little while to recover but is not damaged in any way except that honey production falls off until the complement of flying or foraging bees is built up again.

The same kind of manipulation can be carried out with two unrelated colonies, although it is essential that a good nectar flow is in progress otherwise considerable upheaval will occur. I have had several colonies working on oil seed rape, some of them put there to bolster their growth. Where one colony is larger than another and requires no further stimulation to growth, I have swapped its position with a smaller one. As the larger colony has a proportionally greater number of flying bees they increase the potential of the smaller colony because they return to the hive now occupying the position of their former home. As the flying bees are loaded with nectar and pollen they will be received into the other hive in quite a peaceable manner.

8 Autumn and Winter

Most of us regard autumn as a beautiful season of the year and it is a particularly satisfying time for the beekeeper. Nucleus colonies will have grown to maturity, new queens will have been raised, swarms collected and numerous little problems encountered through the year will have been, we hope, successfully solved. The apprehension of early spring will have been forgotten, and sometimes there is a sigh of relief that the more hectic activity of summer is over. Most satisfying of all, there is honey to be harvested. When that is completed, it leaves only the careful preparation for winter until the seasons change and the whole cycle begins again.

Clearing the supers

Near the end of the summer most colonies of bees will seem a little irritable. With winter approaching, an almost desperate urge seizes the bees to store as much as possible. Robbing becomes more of a problem, not only from other bees seeking to augment their stores but also from intruders such as wasps. It is little wonder that when the beekeeper starts pulling a hive apart to examine the contents the bees react unfavourably to the biggest robber of all. So do not be surprised if those delightful insects you have carefully looked after all year seem reluctant to part with their honey stores.

Taking off the supers of honey is carried out during a period lasting from about the second week of August to the end of that month. In areas where there is heather honey to be collected it will be a little later. The only practical way to take off supers is first to empty them of bees. A second crownboard is fitted between the brood chamber and supers with one or more bee escapes in place. It can be fitted in place of the queen excluder or placed over it. It is my practice to leave the queen excluder in place as I use a board with a sliding door, designed to let the bees pass back into the supers and clean them out after the honey has been extracted. There is only a slight chance that the queen will pass through into the supers but there is no use in tempting providence. Perhaps one day, for peace of mind, I will fit a piece of queen excluder over the side of the access hole opposite the sliding door. I should then be able to dispense with the queen excluder while clearing the supers; it is sometimes a hindrance to the bees, depending on where the bee escape lines up over the excluder.

If, after forty-eight hours, the supers are not empty of all but one or two stragglers then it is possible that one of the bee escapes is incorrectly fitted or, if the Porter-type is being used, the spring strips are not adjusted properly or are gummed up with propolis. If this is not the case, check the fit of the supers to the brood box and to each other. I once had a colony with three supers full of capped honey, possibly 70lb (30kg) or more when extracted. The crownboard and bee escape was fitted, and I returned a few days later to find a regular traffic of bees to and from a small hole in one super. It was where a joint had sprung apart and was just wide enough in one place, about ½in (13mm), to allow a bee to pass through. Needless to say, the supers no longer contained 70lb or more of honey! This stealing of honey seemed to be conducted exclusively by the bees in that hive, transferring it into the brood box. All the better for their winter stores, but hardly the object of the exercise as far as I was concerned. Strangely, there had been no obvious robbing by other bees.

Do not leave the supers empty of bees on the hive. If you do, it is surprising how many other pests will get in, even where the bees find it impossible. Earwigs are always about a hive but do not appear to do any harm, but the same cannot be said of the wax moth or of wasps. The common wasp seems remarkably adept at finding its way into empty supers; the queen wasp likes them particularly for winter hibernation, and to build a nest in during the spring.

After the supers have been cleared there will seem to be an excess of bees in the brood box. A reasonably strong colony may have bees clinging around the entrance as the last of them return in the evening. There is always a little more room after the drones have been evicted, but there must be a higher than normal death rate amongst worker bees in these conditions. It is probably for the best as the strongest will survive thereby ensuring that the overwintered colony is in the strongest possible condition.

When the supers have been removed and the honey extracted, they can be replaced on the hive for the bees to clean. It is surprising how much honey clings to the combs. A layer of honey sufficient to 'wet' the combs will add up to several pounds weight. The common hand-operated honey extractor cannot cope with this residue and it is best cleaned from the combs before storing them for winter. So it makes sense to feed this honey back to the bees.

This is where a crownboard fitted with the small sliding door arrangement mentioned earlier is useful. The supers can be replaced (complete with crownboard) onto the board fitted over the brood box. The sliding door is opened, often with some difficulty as the bees try and gum it up with propolis, and the bees let through into the supers. Two or three days is sufficient for the bees to clean them out. The bees do not re-store the honey in the supers but transfer it to the brood chamber, rather as they did when I had the faulty super fitted. The sliding door is then moved to the closed position and the bees leave the supers through the bee escapes.

Bees can be persuaded to leave the supers by means of chemical repellents. One such repellent which has gained favour in recent years is benzaldehyde. I have yet to use this method of clearing but an associate of mine uses it almost exclusively.

Benzaldehyde is a liquid and a small amount should be sprinkled onto a cloth stapled onto one side of a crownboard. The crownboard on the hive is removed and the bees around the top of the frames smoked down between them. The treated crownboard is placed cloth face down over the frames. A few minutes later the super should be clear.

It is best to treat supers individually, removing each cleared one aside. My colleague had some interesting reactions from his bees when he first tried benzaldehyde. He splashed onto a cloth what he estimated as a sufficient amount to clear three supers at one time—it was almost sufficient to force the bees to vacate the hive! As he watched, they came pouring out of the entrance to escape the repellent. It was fortunate that he is a man of quick reactions. It apparently took him about three seconds to remove the treated crownboard and the supers to let the bees get some fresh air. Fortunately, they survived this treatment and were working well the next summer.

Checking the colony

To survive the winter a colony of bees must be in the very best condition, free from pests and diseases, and it must contain a healthy queen and have plenty of food to hand. A pre-winter check to make sure everything is in order is well worth while. This is best done after the supers have been removed and before the winter feeding starts. Once the bees have been fed, it is best not to disturb them.

Do not expect to find masses of brood and freshly laid eggs. As the income of food dwindles so the queen's laying rate will slow, and eventually she finishes laying until the following spring. Consequently, the brood nest will contract to perhaps half the size it attained during the height of the summer.

Check for the queen; if you have previously coloured-coded her, she should not be difficult to find. If an unmarked queen is found to be present you will know that the original incumbent has been superseded. If this is the case, mark the new queen with the appropriate colour for the year. As long as the bees appear to be disease-free, there is little else that can be done, except to ensure that the hive is in sound order and that plenty of food is supplied for the winter ahead.

97

If there are problems such as severe weakening of the colony caused by disease or an infestation such as acarine, it can be treated but it may be too late to save the bees and kinder in the longer term to destroy them. A queenless colony can always be united with another, but it is usually not possible this late in the year to rear, or even acquire, a spare queen. Similarly a drone-laying queen, which is easily discernible by the type and pattern of the cells produced, can be culled and that colony united with a sound colony. It is very much a tidying-up operation intended to give the bees the best chance of survival since only the strong will live to see another season.

Preparation for winter

It is necessary to ensure that the hives are weatherproof and that they will provide a sound, dry, draught-free home during the forthcoming winter. I like to see all the British National hives in our apiary changed for hives that have been reproofed with preservative. Using the wood preservative

Overwintering WBC hives. Well tied together and sheltered from the prevailing wind by the adjacent hedge. What at first appears to be an untidy grouping is a deliberate arrangement to avoid drifting *(author)*

Cuprinol has the advantage that a hive can be treated and put into use the next day.

This hive changing is done just after the final removal of the supers and it is an ideal time to check each colony thoroughly. It is not necessary to have a full set of duplicate hives as a few can be replaced at a time and those emptied hives reproofed, cleaned and used to replace others. Anyone, though, with perhaps twenty or more hives and only one spare would find such a system impractical and if the hives were not interchangeable life could become very difficult.

Double-walled hives are nearly always weatherproof even when the outer lifts are not in the best condition. It is rare to find the inner chambers of a WBC hive suffering from damp unless the roof is damaged and leaky, then the beekeeper has only him or herself to blame.

So, do whatever you can to ensure that the bees are housed in hives suitable to withstand the ravages of winter and that the hives are tucked away in neither an exposed position nor one susceptible to damp. Hives kept in the garden or those on a permanent apiary site should have already been positioned with some care.

Feeding is the next requirement. Each full-sized colony of bees, occupying ten or eleven British Standard deep brood frames,

A hive fitted with a Miller type feeder. When filled with sugar syrup, the crownboard is placed over it to prevent robbing. With the type of crownboard shown, the slot for fitting a bee escape must be blocked off. Points to note: the entrance block fitted to give the minimum opening; bricks, which make a useful substitute hive stand; on this particular hive the floor is in poor condition, which is something to look out for when purchasing secondhand *(Jak Photography)*

will require 20–25lb (9–11kg) of sugar made into a thick warm syrup. The intention is to make sure the bees are fed sufficient during September to last them through to April. Sugar candy does have its uses in cases of genuine emergency but one should normally discount its use. Use of candy on a regular basis is an admission that insufficient stores were fed to the bees, a situation for which there is no excuse.

One or two hives in the garden or a few in a small apiary can be fed and made ready for winter by the end of the second week in September. In our apiary we use Miller-type feeders almost exclusively that hold approximately 2 gallons (9 litres) of syrup. These feeders will often be emptied within forty-eight hours—it is really surprising how quickly the bees empty them and where they put all that syrup.

This type of feeder, which covers the whole of the brood chamber, is placed on the brood box directly over the frames. The crownboard fitted to the hive is then placed on top of the feeder to cover the exposed syrup and defeat the possibility of robbing. When the feeder is in place and filled, it may be necessary to remove the glass cover over the bee access slot and dribble a little syrup down into the brood box. This is because the bees may not find this food straight away as they have to leave the brood nest to search for it. I like to dribble the syrup down the side of the bee access slot to form a trail to lead them to the food supply. Normally, this is not necessary with the larger feeders, but small feeders with restricted access, especially if it is cold and the bees reluctant to leave the brood nest, can cause problems.

The syrup mixture needs to be strong and thick so that it requires only the minimum amount of water evaporated to make it suitable for storing. Dissolve 10lb (4.5kg) of refined white sugar per gallon (4.5 litres) of water. This is almost at saturation point, and the water/sugar mix will need to be heated until all the sugar dissolves. Stir constantly and do not let the sugar burn at the bottom of the pan as this will render it unsuitable for the bees. Do not use unrefined or brown sugar for winter feed as these are harmful to bees.

Dealing with large quantitites of syrup is difficult on the kitchen stove. A small boiler such as that used for boiling clothes is useful, although the mixture need not be boiled to dissolve the sugar. If the syrup is excessively thick the bees seem to find it difficult to cope with and as it cools it may crystallise. Sometimes on a cold day some of the sugar content will crystallise. In a rapid feeder this is not particularly disadvantageous, but a thin layer of crystallised sugar may block a contact feeder. The syrup is best fed to the bees in the evening as this reduces the chances of robbing.

When the required amount of food has been taken down into the brood box, the feeders may be removed and the crownboard replaced or the feeder hole blocked, whichever is applicable depending upon the type of feeder used. The shutters of a double-walled hive can be closed down to about ½in (13mm), and the entrance block of box hives fitted to give the minimum access. This is often best done before feeding.

Mouse guards should be placed across the hive entrance to prevent mice entering the hive and adopting it as a snug winter home. In the active part of the year they would not risk venturing into an occupied hive, but when the colony is inactive and clustered tightly on the frames the bees do not present so much of a threat and mice will take up residence or just enter the hive to scavenge.

Mouse guards can be made from wire or perforated zinc. I find perforated zinc with ⅜in (9.5mm) holes ideal. When I still kept WBC hives, one of the shutters was permanently fitted with a zinc strip. Normally it was fitted on the projecting end, but for winter the shutter was reversed to place the zinc across the entrance. It was thin enough to overlap the mating shutter so the entrance could still be varied with the guard in place.

In some areas of Britain woodpeckers are reported to be a nuisance. They will make a hole in the hive wall and help themselves to a good meal. It seems a fairly localised problem, almost as though the birds learn the trick, but if it is likely to affect you the best prevention is to cover the hives with wire netting. Garden netting, such as that sold to protect soft fruit, can be a deterrent when draped over several hives and pegged down.

Hives should be stripped to the minimum for wintering. That does not mean that they should be in any way flimsy, but all extra lifts and supers should be removed. After all, these wooden boxes (for that is all they are) have to stand out all winter to be subjected to rain, snow, ice and maybe gale force winds. Double-walled hives are particularly susceptible to having their roofs blown off, especially if they are gabled. The traditional remedy is to put a couple of housebricks on top, but this is not as secure an arrangement as it may look. It is far better to rope the whole assembly together with nylon cord. Box hives are less likely to have their roofs lifted off by the wind, especially if a very deep roof is fitted. Even so, most box hives in an apiary are seen topped off with a housebrick during the winter months.

Winter

This is the quietest time of year for the beekeeper, and it is useful to spend some of that time profitably in the workshop. There will be equipment to be repaired and stored away until the following season. I do not believe in leaving anything out of doors that can be put in the garage or garden shed or even under a simple veranda-like shelter. Equipment such as extractors should always be kept indoors.

Supers containing extracted comb must be stored in such a manner that they are protected from mice, hibernating queen wasps and, worst of all, the wax moth. It is usually convenient to stack supers in a corner. They should be off the floor; a blocked off crownboard placed on four housebricks or blocks of wood is a good support. The individual supers can be separated with sheets of newspaper and a mothball placed in each separate super to deter the wax moth. Supers so treated will, of course, need a good airing in the spring before use.

The bees are not as inactive in winter as you may think. If the temperature rises sufficiently, a few may even leave the hive on cleansing flights. Another sign that they are active is the presence of dead bees and rubbish pushed out of the hive. This is always much more noticeable on hives fitted with a flight board. The dead bees will be pushed just clear of the entrance and accumulate on the flight board until either the rain washes them off or the sparrows eat them.

Without a doubt, the amateur with his or her first colony will be apprehensive as to whether the bees have sufficient food for the winter. In Britain feeding candy on Boxing Day is still something of a ritual and a talking point. So the beginner who has carefully checked the bees and fed them with perhaps 25lb (11kg) of sugar in syrup form will almost feel guilty when asked 'Have you put the candy on yet?'. When this question is asked by a resolute individual of many years' beekeeping experience, it is sufficient to induce at least a twinge of panic. So, you will want to make quite sure the bees are well provided for and give them a block of candy. Do it if you must; it will not harm the bees, but when your confidence increases you will do well to dispense with it as unnecessary. If you do need to use it, try the following recipe.

Dissolve 1lb (450g) of sugar into $\frac{1}{4}$ pint (14cl) of water, stirring well. Bring to the boil and add one teaspoonful of lemon juice. Boil for two minutes after all the sugar has dissolved. Place the pan containing the sugar/water mix into a shallow container of cold water. As the mixture begins to become cloudy, stir well and pour into a suitable receptacle. Leave to solidify. When solid, the candy may be placed over the feeder hole on a hive. A plastic 1lb (450g) margarine tub placed over the candy will prevent the bees leaving the hive when they have eaten through the candy. It also means the addition of a shallow super to the hive in order to give sufficient space below the roof for the block of sugar candy.

9 The Harvest

Honey bees are wonderful pollinators and quite fascinating insects to study, but primarily one keeps them to produce honey. This may only be intended for home consumption but quite often there is sufficient surplus to sell a little and show a modest profit. A visit to the local delicatessen or health food shop makes one quickly aware that packaging is an important aspect of marketing honey. It is not really possible to sell honey loose, although there is no harm in selling it to friends and relatives in jam jars, but greater care is necessary when selling on the open market, either at the door or to shops. Whichever way it is sold, some additional expense is involved. It may be extracted and sold in ½lb (225g) or 1lb (450g) jars or larger honey tins, or sold in comb form as sections or cut comb.

The use of sections is attractive to the beekeeper with one or two stocks who does not wish to burden him or herself with the paraphernalia of extraction equipment. However, the use of sections is not without its problems, and there is fortunately an easier method of producing saleable goods without resorting to honey extraction. This is achieved by producing conventional supers of comb, albeit on thin unwired foundation, and cutting it into blocks which can be sold in plastic tubs as cut comb.

If honey is sold in extracted form there is the cost of jar, lid and label to add to the basic price per pound. On a small scale this cost can be partly offset by asking for jars to be returned when sold to individuals and by typing or writing one's own labels. Sections are an expensive way of selling honey. Not only is the cost of the wooden or plastic section and section case to be taken into account, but also the loss of drawn out comb. This is also applicable to cut comb but some saving is made on packaging by selling it in plastic tubs with 'snap-on' lids or even

transparent bags. All these types of packaging are available from the larger suppliers of beekeeping equipment.

It should be noted that the sale of honey is subject to several legal requirements. A guide to these regulations appears at the end of this book.

Sections

When sections are produced you will find that, unless you are very lucky, a number of them are not suitable for general sale and must be kept for home use. The normal faults are that the comb is not fully drawn out or not completely capped. Propolis may make a mess of sections but can easily be cleaned off the new plastic section boxes or gently scraped off the wooden ones. If the sections have been drawn out quickly in a good honey flow they should be free from propolis. With a poor or intermittent nectar flow the sections will take several weeks to be filled and then the wax cappings become rather yellowish instead of a fresh white and do not have an attractive appearance. If some of the production is for the home, the substandard goods can be used up there and the best kept for sale.

Sections are sold in card boxes with a transparent window. This box or case is purchased flat and then folded to fit around the section. In this way the goods can be seen in the box and it does provide a visually neat appearance. Transparent bags made especially to hold sections are now available, but while more economic to purchase they do not make such an effective package as the card boxes.

It is particularly useful to produce honey in comb form when the bees are working ling heather. This is a very viscous honey and has a peculiar property known as thixotropy; this means it is a natural jelly which, when agitated, becomes fluid and then returns to jelly form after settling. It therefore cannot be extracted from comb in the normal manner and is usually pressed out which unfortunately destroys the comb. So, it can be an advantage to consider producing heather honey in sections or cut comb form.

Cut comb

This is the alternative method of marketing comb honey and it is more economical than producing sections. There is still the same loss of drawn out wax comb to be taken into account but some saving is made on the packaging. It also has the advantage that blocks of comb may be cut to different sizes and weights, thereby giving more flexibility to price and marketing technique. One use of cut comb is to cut a piece square in section, ie the same width as the thickness through the comb, and about the same length as the depth of a honey jar. It is sold in the jar which is filled with liquid honey and very effective it looks too, but a bit messy to do anything with.

The more common method of presenting cut comb is in plastic tubs with 'snap-on' transparent lids. The comb can be cut freehand to fit but must comply with a minimum weight shown on a label which usually carries details of the vendor's name and address. It is possible to purchase a comb cutter manufactured to cut comb out of a frame and of a suitable size for the container. When cutting freehand, however, it is best to use a large kitchen knife and warm it, before use, over the stove—not on the stove, as it only need feel warm to the touch. This will greatly aid the passage of the knife through the comb. Remember that this is a special thin unwired foundation—the recipient of a piece of wired comb may not understand and feel somewhat less than delighted!

A British Standard shallow comb from a super frame cuts into six or seven pieces approximately $2\frac{1}{4} \times 3\frac{1}{2}$in ($5 \times 8$cm), each piece weighing approximately $\frac{1}{2}$lb (225g). I generally find it easier to cut the comb vertically downwards while holding the frame away from me at a slight angle. Then it is an easy matter to cut around the inside of the frame releasing the separate segments.

The disadvantage of first cutting around the frame is that the whole block of comb will fall out and then has to be cut up with one side flat on a plate. This results in bruising or breaking open of the capping on the underside. This side will probably be placed face down in the container but to my mind it makes the job less neat than it should be.

The separate blocks of comb may be drained of surplus honey where the cells have been cut on the outside. This can be done by placing them carefully on a thin metal cake rack and allowing the honey to drain onto a large plate or tray. After a few hours it is a simple matter to pop them into the cut comb container and snap the lid into place complete with sticky label bearing all necessary details.

Extracting and bottling

Extracting honey causes the least waste as one retains the comb for future use and the wax cappings may also be reused or even sold. This method does have disadvantages, of course, with extra equipment, complication and time. As a minimum one needs the following equipment or some substitute equipment: extractor; uncapping knife and tray; honey tank; honey jars with lids, seals and labels.

There is no effective substitute for the extractor, although there is a hand-held device now being sold which resembles an electric motor with a clamping arrangement for holding single frames. It is not an extractor in the conventional sense but none the less the principle is the same. Usually the loan of an extractor can be arranged through a local beekeeping organisation or sometimes from a school or college.

A heated uncapping knife and tray are luxuries I have always promised myself but, like many amateurs, I soldier on with a large kitchen knife, an aluminium tray and a container of hot water. For several years I did not have a honey tank (sometimes called a settling tank or ripener) and then I obtained one secondhand. I count it amongst the most useful items that I have. It is a tall cylindrical tank with a honey tap at the bottom and a false section at the top with a perforated screen which acts as a coarse filter.

103

Cutting cappings from a super full of honey; this should, ideally, be done with a *downwards* movement. The bowl contains hot water to warm a spare knife in preparation for cutting the second side of the comb. The honey here will be extracted with great difficulty, as it has started to granulate due to a high content of nectar from oil seed rape (*Jak Photography*)

Honey jars, lids, seals and labels can only be dispensed with if one is selling in bulk either in 28lb (12kg) capacity tins or polythene buckets, but most of us like to see honey in jars. Fifty or sixty or more lined up on the kitchen table, the culmination of the year's work, is a sight worth seeing, especially if it represents the first year's produce.

The actual extraction of honey does not present any problems as long as it is tackled in a methodical manner. The first move is to decap the comb in the shallow super frames. There is an air gap between the cappings and the honey, and ideally this is where one should slice the cappings off to expose the honey. In practice, it does not work quite like this as the bees do not draw out perfectly flat comb. The most perfect frames are those outer frames with one side to the hive wall. These are drawn out very uniformly and appear almost flat.

I decap comb with a large kitchen knife or, to be more precise, two of them. They are fairly sturdy, which is useful to retain the heat, and long enough to cut the cappings off a British Standard deep brood frame of comb if necessary. On the kitchen stove I have a container of hot water which is kept just below boiling point. Into this the two knives are placed, then withdrawn and replaced alternately as the operation progresses. The idea is to have to hand a warm clean knife for each frame of comb.

The frame of comb should be held on a tray or other suitable shallow receptacle for catching cappings and some dribbles of honey. A large plate is not ideal as the end of the frame resting on it will tend to slide about. Not only is this inconvenient but, when using a razor sharp kitchen knife, it could be dangerous. The aluminium tray which I use is homemade of welded sheet construction about 18 × 12in (45 × 30cm) with a flat strip across the middle. This strip has a 1¼in (30mm) hole at about its centre point and the lug of the frame being

A great rarity—one of the first honey extractors, Abbott's 'Little Wonder', c1875, at Pershore Agricultural College (*Jak Photography, by kind permission of Jim Crundwell*)

decapped is slotted into it for a firm hold.

The frame is held at a slight angle and one of the preheated knives passed down its full length to slice off the cappings. Where the comb undulates do not bother to fiddle around: cut cleanly across the full face. The honey which runs off or is cut away can be drained from the cappings afterwards. Cut *down* the frame; this is not only much safer than cutting upwards and towards oneself but also has the advantage that loose cappings will fall away from the knife into the tray.

On a small scale frames of comb can be decapped and placed straight into the extractor. Extractors are available in a variety of sizes with the frames arranged tangentially or radially and with manual belt, chain or gear drive or electric power. Most home enthusiasts make do with a manually

MG wax extractor in use in the home; on this particular example the funnel from the large spout is missing (*author*)

driven machine which is quite adequate for the task. Many of the smaller manual extractors are arranged to hold the combs tangentially. Because of the strain this places on the comb, fine wire grids are fitted into the main framework and it is against these grids that the comb is placed. With these mesh grids in place one can extract at quite a rate, although it is advisable to balance the extractor by placing frames of approximately equal weight on opposite sides of the holder.

If you purchase an old extractor and the grids are missing, it can still be used with a little care. It cannot be operated quite as vigorously as one fitted with grids and it is necessary to extract partly one side of the comb and then change it around to extract the other side. If this is not done it is most likely that the weight of the honey on the

inside face of the comb will, during extraction, cause the comb to collapse and fly to pieces.

As extraction proceeds, the level of honey in the bottom of the extractor will build up until it starts to drag against the projecting lugs of the frames being extracted: that, of course, is assuming a reasonable harvest. Most of the honey should be then drained off, preferably straight into a settling tank.

When extraction is finished and the extractor drained of honey, which may take all night, it should be thoroughly cleaned especially if it is made of tinned sheet steel. Hot soapy water will get rid of any residual honey and wax. A soft-bristled rubber-backed brush such as used for car cleaning is very useful for cleaning out an extractor, especially the rotary frame.

After a couple of days the honey in the settling tank will have cleared. The bulk of the loose pieces of wax will be trapped in the

Small honey extractor. Frames can be fitted tangentially and semi-radially in this type of extractor (*Jak Photography*)

top section by the perforated screen. Any wax and other debris, such as bits of bees and lumps of pollen, will have floated to the top in the main part of the tank. The honey to be bottled is drawn off from the tap at the base of the tank. This way, most of the bottled honey will be clear and bright and look very presentable. The honey which is drawn off last and contains all the debris is not worth bottling except for home consumption, and can be stored in bulk in large margarine tubs or jam jars. There is nothing wrong with this honey/wax/pollen/bee mixture and it can be used for cooking instead of white sugar. With care, and by running the honey through an extra screen which can be fitted under the tap on the extractor, there will not be a large amount of this 'impure' honey. Apart from cooking, it can be used for the table. Health food enthusiasts would no doubt approve, but do not be surprised if the more fastidious members of the family refuse the honey pot when faced with the sight of a bee's leg floating in a morass of wax cappings. Traditionally, the beekeeper always finishes up with the leftovers and the best is offered for sale.

When bottling honey, take care to ensure that the correct minimum weight is in the jar, or even a little over. Some of the new plastic honey containers are marked with a line to denote the minimum quantity but the more traditional wax jars are not so marked. The best thing to do is to weigh a jar complete with lid, then add this weight to the 1lb (450g) of honey. A standard squat jar with lid weighs approximately 7oz (175g). If this is filled with honey until the total weight is 1lb 7oz (625g), then we are sure the customer will be getting a fair deal. Once one jar has been done by weight, note the level of honey and use it as a master to check the other jars against.

When all the jars are filled, wipe them over with a warm damp cloth to rid them of any stickiness and affix the labels. Labels to a number of designs can be purchased from suppliers of beekeeping equipment and often arrangements can be made to have one's own name and address overprinted. If this seems a little extravagant, a name/address rubber stamp can be purchased from a stationers and proprietary labels stamped with ink. Your own name and address on the label is all good advertising and, after a season's work, something to be proud of.

Wax extraction

At the end of the season one is usually faced with a pile of wax cappings, pieces of old comb and often odd bits of natural comb which in themselves are of little value, although beeswax in a useful form is of value. What is needed is a method of wax extraction which will leave behind all the impurities and allow the beekeeper to either sell or reuse the pure wax.

There are several methods of achieving this: the solar extractor, the MG hydraulic extractor or various steam extractors. The larger steam extractors are undoubtedly useful on a semi-commercial scale but an expensive extravagance for the average amateur. The MG hydraulic extractor works fairly well but is a little messy in the kitchen. The extracted wax is not particularly pure and the device can become blocked if a large amount of really black old comb is put through it.

The simplest type is the solar extractor which will cope with any old comb and bits and pieces. This is rather like a tray with two sheets of glass in a framework which acts as a door. The whole assembly is set up at an angle out of doors facing a southerly direction. The old comb and cappings are placed above a weir plate which is fixed towards the lower end of the extractor. The warmth from the sun is magnified by the glass double glazing and the wax melts and runs down over the weir leaving the rubbish behind. As the wax runs into the lower section, it is channelled away into a receptacle.

Wax obtained by any of these methods can be reheated and poured into moulds of a known capacity. The moulds should be swilled out with a little soapy water to prevent the wax adhering to them. Do not apply direct heat from a stove to a container of wax but melt it by an indirect method such as placing the container in a pan of boiling water. This works well with two saucepans, especially if the smaller one containing the wax has a lip designed for pouring.

So, very little is wasted, the best honey is sold, the 'seconds' used in the home and the wax either sold, used for polish or perhaps hoarded until you have sufficient to try your hand with a small press made to produce foundation. Even propolis has a market, albeit somewhat limited, and you should look out for advertisements requesting supplies. Pollen also enjoys a certain demand and is used in health food shops as part of some preparations. I have seen 1lb (450g) honey jars containing cappings being sold at nearly half the price of 1lb of honey in such establishments!

10 Pests and Bee Diseases

Pests and diseases are facts that cannot be ignored, but there is nothing more disheartening to the beginner than to be faced with an extensive catalogue of gloom early in his or her apprenticeship. The newcomer to beekeeping should therefore be aware of the possible problems, preventative measures and cures, but should keep them in perspective.

The keeping of any form of livestock, whether it be the family cat or a pedigree herd of cattle, is subject to the problems of illness and disease, and bees are no exception. Should your bees exhibit one of several diseases, it is a matter of concern but nothing to be ashamed about. Healthy bees are vulnerable to disease in the same way that robust healthy humans are. One of the most important lessons for the beginner is: if in doubt, ask. It is far better to approach someone more knowledgeable when you think that there is something wrong with your bees, than to be faced with a rapidly dwindling colony which may become too weak to survive. The old adage, prevention is better than cure, should be adhered to whenever possible.

There are diseases which affect adult bees and others which affect brood. It is easier to identify diseased brood, or at least to become aware that something is amiss, because it is static in the comb and therefore easier to examine. A few unhealthy adult bees go unnoticed amongst the thousands in a hive and do not cause any problems until a great many others are affected.

Pests

Greater wax moth (*Galleria mellonella*)

Dull brown and ash white in colour, wing span 1–1½in (25–38mm), edges of wings appear ragged.

Nature of problem
The wax moth lays eggs amongst wax comb, either in store or in an active hive having a preference for brood comb. The larva devours the comb, burrowing through it and leaving what appear to be silken tunnels.

On preparing for pupation the larva will tuck itself away, quite often between the frame top bar and hive walls. It forms a boat-shaped trough in the wood before spinning a cocoon which can cause considerable unsightly damage.

Prevention
See lesser wax moth.

Lesser wax moth (*Achroia grisella*)

Silvery in colour, wing span $\frac{5}{8}$–$\frac{7}{8}$in (15–20mm), edges of wings appear ragged.

Nature of problem
Unlike the greater wax moth, the lesser wax moth will not live amongst the comb in a hive, but the damage caused by the larva is similar. There is no damage to the woodwork on pupation but the effects of the silken tunnels are more extensive than those of the greater wax moth.

Prevention
It is virtually impossible to prevent either species of wax moth entering a hive to lay eggs. The best time for observing them is the evening when they may be found clinging to the side of the hive or up under the roof. A strong colony of bees will usually deter the wax moth, but a small colony with surplus combs may well be subject to attack. Colonies kept in large glasshouses or polythene tunnels for pollination purposes rapidly weaken and are also particularly susceptible to excessive wax moth infiltration. I have seen a hive where the wax

moths and larvae occupied more of the hive than the bees.

The only measures that can be taken against the wax moth in the hive are: (*a*) to ensure only the minimum amount of underutilised frames are fitted (maximum two); (*b*) to destroy wax moths at every opportunity; (*c*) to cut out and destroy the larvae when found in comb.

Stored comb, such as supers at the end of the season, can be protected with ordinary household mothballs. The complete supers containing the frames of wax should be stacked together with a sheet of newspaper separating every other one. Part of a mothball may be placed on each sheet of newspaper to protect the comb placed above it. It is most important that the supers and contents are well aired before use on a hive. I like to have them opened out on the lawn for a couple of sunny days; this is usually sufficient to clear the smell.

Wax foundation can be protected by keeping it in sealed polythene bags. If it is stored, well-wrapped, in a cupboard in the house it should be free from the attentions of the wax moth.

Bee louse (*Braula coeca*)

Small pinhead-sized red mite. A member of the fly family that has lost its wings during the evolutionary process.

Nature of problem
The only effect the bee louse has on a colony is damage caused to cappings on sealed stores. The eggs are laid in the cappings and the larvae tunnel along through the capping which can weaken it and let honey ooze out.

The bee louse is most often seen riding on the backs of the bees, especially the queen. They do not appear to cause harm or damage to the bee in any way.

Prevention
There is no known way of ridding a colony of the bee louse, or of stopping them entering the hive. It does not seem to matter as they appear harmless.

Pollen mite (*Family: Tyroglyphidae*)

Small hairy mites found in stored pollen. Not visible with the naked eye.

Nature of problem
Pollen mites are beneficial rather than harmful, clearing away patches of mouldy pollen by turning it from a hard block to powder which is readily removed by the bees.

Prevention
Not possible or necessary.

Varroa (*Varroa jacobsoni*)

A very small mite which resembles a tiny crab in form. They are a different shape from the bee louse, *Braula coeca*, and belong to the family of spiders, having eight legs, not six like the bee louse. The female is oval in form and brown in colour and the male, which is rarely seen, is more circular in shape with a greyish white or cream colouring. The female is about $\frac{1}{25}$in (1mm) long by $\frac{1}{16}$in (1.5mm) wide or slightly larger.

Nature of problem
Varroa is a true parasite, clinging to the bees, often on the underside of the abdomen partly hidden by the overlapping segments. They pierce the intersegmental membrane and literally live off the unfortunate bee. The eggs may be laid on bee larvae and therefore become sealed in the cells with the developing bees.

Bees infested with two or more parasites appear restless as they try to free themselves of the mites. If the mites have attached themselves near the wing roots, flight may be difficult or impossible. Bees thus affected will eventually crawl away from the hive and die. In a severely infested colony the infested bee larvae will die and crippled bees with malformed wings, legs or bodies may be found piled in front of the hive.

Prevention
Fortunately, varroasis, the term used to describe an infestation, is not yet known in Britain. According to the Ministry of Agriculture, it can only be introduced into this country by the import of affected bees carrying live *Varroa jacobsoni* mites. In countries where varroa is present it has spread with alarming rapidity and caused enormous losses of bee colonies.

111

There is no known prevention or cure but it is most useful for all beekeepers to take part in a diagnostic search for this mite. The National Beekeeping Unit at Luddington Experimental Horticultural Station, Stratford-upon-Avon, Warwicks CV37 9SJ, has issued a leaflet on varroasis which I would urge any beekeeper in Britain to obtain.

Other articles have been prepared by the BBKA and the British Isles Bee Breeders Association. Any of these organisations will be only too willing to assist with any queries concerning this particularly harmful pest.

Adult bee diseases

Acarine (formerly known as Isle of Wight disease)

Cause
Acarapis woodi, a small mite which lives in the main thoracic trachea (breathing tubes in the thorax) of the bees.

Symptoms
According to the Ministry of Agriculture, there is a small percentage of bees infected with acarine in most hives. A high rate of infestation usually follows a poor summer season when the bees have spent much time confined to the hive. This will result in a high death rate in the winter and during the spring when they start becoming active; piles of dead bees will be thrown out of the hive.

Bees with acarine will crawl up grass stems at the front of the hive and seem unable or disinclined to fly. In the more advanced stages of infestation the whole colony will exhibit a lethargy which is quite different from the normal hectic summer activity. A good number of bees may be observed with their wings carried at a strange angle which gives them an almost hunched, awkward appearance.

Treatment
The best cure and preventative for acarine is the use of Folbex smoke strips. These are manufactured by CIBA-GEIGY Ltd, Switzerland, and are available from all major

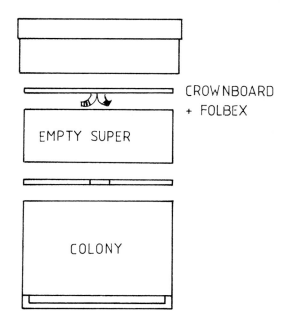

CROWNBOARD
+ FOLBEX

EMPTY SUPER

COLONY

Fig 21 A method of treating a hive with 'Folbex' smoke strips. Note the entrance block in the closed position. The crownboard over the hive is shown with the feeder hole open, to allow the smoke to enter the brood chamber

suppliers of beekeeping equipment. The smoke strip is impregnated with the selective acaride chlorobenzilate.

Each colony should be treated either in the morning or evening. Treatment during the early spring is probably best as this gives the colony time to build up during the forthcoming summer, but if acarine is suspected then it is wise to repeat the treatment in the autumn.

One smoke strip for each hive is sufficient. It should be hung in an empty super placed over the hive with the feeder hole in the hive open and the entrance closed to prevent the bees leaving. It is no use lighting a smoke strip and shoving it into the hive entrance because the strip will not burn where it comes into contact with the hive. If the strip attempts to burn rapidly, the flame should be blown out and then the strip will smoulder slowly.

As the smoke drifts down into the hive the bees will start fanning with quite an audible roar. This has the effect of forcing the smoke into every crevice and out of every crack in the hive. The smoke will kill the active mites and the treatment should be repeated 8–10 days later.

Amoeba

Little is known about amoeba or its effect on the bees, possibly because it is not very common. It is known that sterilisation with acetic acid will kill amoeba.

Dysentery

Cause
A malfunction of the gut, possibly caused by excess water in the winter stores. The presence of honey stores which crystallise rapidly, such as oil seed rape, seem to be a contributory factor.

Symptoms
In the worst condition a colony will die out, and on examination in the spring the hive will be found to be almost glued together with faeces. If the winter is mild, the problem is not so bad as the bees can leave the hive on a sunny day to void the contents of the gut.

Treatment
There is no known treatment, but if each colony is fed with at least 20lb (9kg) of sugar syrup, which is less prone to crystallisation, the chances of dysentery occurring are lessened.

Nosema

Cause
Nosema apis is a single-celled animal like amoeba which lives in the gut of a bee until it is voided in the faeces as spores. If faeces are present on the comb in a hive it will be cleaned up by other bees which in turn become infected. The presence of dysentery considerably aids the spread of nosema.

Symptoms
An infected colony is very slow to build up in the spring because nosema halves the lifespan of an adult bee. By about June the colony has normally recovered but this may affect the chances of a good honey harvest or ruin the prospects of embarking on a breeding programme.

Treatment
Fumidil B should be fed with the sugar syrup winter stores, preferably in a rapid feeder. Fumidil B is a fine powder and difficult to stir into a ready-made syrup so it is best mixed with the sugar before adding the water. It is manufactured by CEVA International, and is distributed in Britain by E. H. Thorne (Beehives) Ltd. It is an antibiotic in the form of a stable formulation of the soluble bicyclohexylammonium salt of fumagillin.

Chronic bee paralysis

Cause
A virus.

Symptoms
The bees literally become paralysed and are incapable of moving even when being smoked. There may well be heaps of dead bees below the entrance of a hive or, with a WBC hive, a number on the flight board being 'worried' by other bees.

This virus seems to have one of two effects on a colony. Either a few bees die out each day which is a small proportion of the breeding rate or, in more rare instances, hundreds are affected and the colony dies out.

Treatment
None is known, but it is believed that there is a genetic susceptibility to this virus and requeening with a different strain often cures the problem.

Brood diseases

American foul brood

Cause
A spore-forming bacterium, *Bacillus larvae*, which is fed as spores in the brood food by the nurse bees to bee larvae. The spores hatch in the gut of the bee larva, remaining dormant until its cell is sealed. At this stage the bacillus breaks out of the stomach into the body cavity, rapidly proliferating and subsequently killing the bee larva.

Symptoms
As the larva rots, it changes from a yellowish

colour to brown and then black as it dries out. At the brown stage the remains will stick to a matchstick poked into a cell and pull out into a slimy, sticky strand.

Cappings of cells containing an infected larva will become sunken and unused by the bees. This results in an ever-increasing number of empty cells scattered about a frame of brood.

Treatment

There is no known treatment when a colony is found to have, or suspected of having, this disease, but the Ministry of Agriculture should be informed so that the local Foul Brood Officer can come and examine the colonies, according to the Foul Brood Diseases of Bees Order 1967.

The best way to destroy an infected colony is to burn the bees, wax and frames. In the evening, when the bees have ceased flying, approximately 1 pint (0.5 litres) of petrol or trichloroethylene should be poured through the feeder hole into the frames. The fumes will quickly kill the bees. The contents of the brood chamber and supers, if fitted, should then be deposited on a ready-made fire. Care must be taken to ensure that all the dead bees are burnt and the hive must be scraped clean of all propolis and wax.

The hive should then be sterilised with a blowlamp or gas torch. Not only the inside, but the outside and the contact faces of the hive must be singed. It is a job better overdone than underdone. The hive may then be reused, but never ever be tempted to try and salvage some of the frames or wax comb from the supers: the risk of reinfection is much too great and such rash action would be regretted.

European foul brood

Cause

A non-spore forming bacterium, *Streptococcus pluton*, fed in the brood food by the nurse bees to the bee larvae. It proliferates in the stomach of the larva feeding on the stomach's contents. Unlike the American foul brood bacillus, *Streptococcus pluton* does not break out of the gut into the body cavity but starves the bee larva of food which will often result in its death.

Symptoms

The larva often dies before the cell is sealed, and the normally shiny white larva will appear yellowish or grey in colour and slump into the bottom of its cell. The sticky ropiness typical of American foul brood does not occur, but the rotting larva may give off an offensive smell.

Treatment

As with American foul brood, a suspected infection should be reported to the local Ministry of Agriculture Foul Brood Officer. It is no longer necessary to destroy colonies with European foul brood; they may be treated by the Ministry Officer with oxytetracycline and the treatment is free on request.

Addled brood

Cause

Brood that dies from defects passed on from the queen. There may be a number of reasons for these deaths which may occur at any time during the brood stage of a bee.

Symptoms

Dead brood.

Treatment

Requeen with a queen from a different strain.

Chalk brood

Cause

A fungus, *Ascosphaera apis*. If the spores of this fungus are eaten by a bee larva they germinate in the gut forming mould-like strands which penetrate the body, resulting in death.

Symptoms

The larvae appear chalk-white, and if pulled out of the cell are found to be quite hard. The nurse bees will drag the mummified corpses from the brood comb and throw them out. On a hive fitted with a flight board the remains will sometimes be seen when chalk brood is present.

Treatment

The colony should be requeened and the frames changed as soon as possible.

Stone brood

Cause
A fungus, *Aspergillus*, which acts on a bee larva in a similar manner to chalk brood.

Symptoms
Unlike chalk brood, stone brood changes the larva or pupa into a yellowish granulation. It is fortunate that this disease is rare in Britain as the fungus *Aspergillus* can cause a disease of the windpipe in man and birds known as aspergillosis.

Treatment
No known remedy.

Sac brood

Cause
A virus which causes death to the larva after the cell is sealed.

Symptoms
The larva dies with its head turned up in the entrance of the cell. This may be visible after the nursery bees have removed the cell capping. Because of its peculiar form, this is often referred to as the 'chinese slipper' stage.

Treatment
Requeen with a queen of a different strain.

Appendix A

SUPPLIERS OF EQUIPMENT

GENERAL SUPPLIERS

Robert Lee (Bee Supplies) Ltd,
Beehive Works,
George Street,
Uxbridge,
Middx

Taylors of Welwyn,
Beehive Works,
Welwyn Garden City,
Herts

E. H. Thorne (Beehives) Ltd,
Beehive Works,
Wragby,
Lincs

Steele & Brodie,
Beehive Works,
Wormit,
Newport on Tay,
Fife

SPECIALIST SUPPLIERS

Brinsea Products,
West Brinsea Farm,
Congresbury,
Avon
*Polyurethane foam hives, honey
extractors, foundation presses*

Foster Magneto Co Ltd,
27 High Street,
South Norwood,
London, SE25 6HA
Plastic frames

Four Acres Hives,
Lambrook House,
8 Lambrook Street,
Glastonbury,
Somerset
Hives

French, Flint and Ormco Ltd,
61 St Thomas Street,
London, SE1 3QX
Honey jars and caps

John Gower, Hivemaker,
Passage Road,
Saul,
Glos
Hives

M. Holden,
Newstead Honey Farm,
East Chiltington,
Lewes,
Sussex
Hives

Leaf Products,
24 Acton Road,
Long Eaton,
Nottingham, NG10 1FR
Plastic foundation, frames etc

Maisemore Apiaries,
Old Road,
Maisemore,
Gloucester, GL2 8HT
Hives, frames, bees etc

C. Mansfield,
21 Encombe,
Sandgate, Folkestone,
Kent
Conical bee escapes

Morris Beekeeping Supplies,
Chaddesden,
Morants Court Road,
Dunton Green,
Sevenoaks,
Kent
Pure beeswax foundation

Mountain Grey Apiaries,
19 Westfield Avenue,
Goole,
North Humberside, DN14 6JY
Bees and equipment

B. J. Sherriff,
'Five Pines',
Mylor Downs,
Mylor,
Falmouth,
Cornwall
Beekeepers suits

South Coast Honey Farms,
42 Priory Road,
West Moors,
Wimborne,
Dorset, BH22 0AY
Hives, frames, foundation etc

Appendix B

REGULATIONS RELATING TO
THE SALE OF HONEY

1. Introduction

There are several legal requirements relating to the sale of honey; the principal ones are:

(i) Weights and Measures Act 1963 (Honey) Order 1977 S.I. No 558
(ii) Weights and Measures (Marking) Regs (Amended) 1977 S.I. No 1683
(iii) The Honey Regulations 1976 S.I. No 1832
(iv) The Food Hygiene (General) Regulations 1970 S.I. No 1172
(v) The Labelling of Food Regulations 1970

2. Weight

The Act requires honey to be marked with net weight and to be pre-packed only in fixed quantities ie 1, 2, 4, 8 or 12oz, 1lb, 1½lb and multiples of 1lb. Containers in which comb honey and chunk honey are made up are required to be marked with an indication of quantity by net weight; hitherto cut comb containers have not been required to be marked with net weight and chunk honey with net weight or capacity. Small quantities are no longer exempt from the requirement to indicate the quantity on the container.

Marking
Containers will be required to be marked with metric units only (although they may be marked with imperial units in addition) after 31 December 1979.

The Honey Regulations
These regulations, which came into force on 1 May 1977, implement the EEC directive on the harmonisation of the Laws of Member States relating to honey. They prescribe definitions for honey, including definitions of blossom honey, honeydew honey, comb honey, chunk honey, drained honey, extracted honey and pressed honey. Water content must not exceed 23 per cent in the case of heather (Calluna) honey and clover (Trifolium) honey or 21 per cent for other honey. Honey must not have foreign tastes or odours or be fermenting or be heated to such an extent as to destroy its natural enzymes. The latter is measured by the diastase activity which must be not less than 4 and the hydroxymethylfurfural content shall not exceed 80mg per kg. It is in order to ascribe the floral source of honey provided it is derived wholly or mainly from the plant indicated and to use topographical or regional descriptions provided that the honey originated wholly in the region indicated. Labels should be clear, legible and indelible and state whether the honey is comb honey or chunk honey for example.

3. Hygiene

These regulations were made in 1970 under powers conferred by the Food and Drugs Act 1955. They underline the need for cleanliness in the preparation of foods. There is a section relating to the use of domestic premises.

4. Labelling

A honey label should always contain the following information:

(i) Description, ie honey, cut comb honey, chunk honey, Worcestershire honey, heather honey etc
(ii) Name and address of the packer
(iii) Weight in metric with the imperial equivalent

Bibliography

General

Bielby, W. B. *Home Honey Production* (E.P. Publishing Ltd, 1977)

Hooper, W. E. J. *Guide to Bees and Honey* (Blandford Press, 1976)

Mace, H. *The Complete Handbook of Beekeeping* (Ward Lock, 1976)

Ministry of Agriculture, Fisheries and Food *Bulletins* (HMSO, London)

Vernon, F. *Beekeeping—Teach Yourself Books* (Hodder and Stoughton, 1976)

Waine, Dr A. C. *Background to Beekeeping* (Bee Books New and Old, 1975)

Wedmore, E. B. *A Manual of Beekeeping* (originally published 1946; London Facsimile reprint, Warminster 1975)

The Ministry of Agriculture, Fisheries and Food (Publications), Lion House, Willowburn Estate, Alnwick, Northumberland, will also supply useful leaflets on all aspects of beekeeping, free of charge.

Monthly publications

Bee Craft The Official Journal of the British Beekeepers Association
Mrs S. White (Secretary),
15 West Way,
Copthorne Bank,
Crawley,
Sussex

Beekeepers News
E. H. Thorne (Beehives) Ltd,
Beehive Works,
Wragby,
Lincs

Bee World
International Bee Research Association
Hill House,
Gerrards Cross,
Bucks

British Bee Journal
46 Queen Street,
Geddington,
Kettering,
Northants

Acknowledgements

My thanks go to all those who have assisted in the production of this book, particularly Sue Hall who was very patient when I did not keep to all the deadlines.

I would also like to thank the following people for their advice, assistance, or simply very welcome encouragement during the compilation of the manuscript: Mr J. S. Crundwell, Pershore College of Horticulture; Mr J. S. Cox, Gloucestershire College of Agriculture; Mr J. Tinson, Leaf Products; Mr L. L. Thorne, E. H. Thorne (Beehives) Ltd, and Mr K. Showler, International Bee Research Association. Also Bob Langston, Colin Musgrove, Mike Sessions, Bill Walker and Mr and Mrs Crowther.

My very special thanks are due to my good friend Sue Walker who converted my handwritten scrawl into a legible typewritten manuscript; without her help this book may never have been written.

Index

Numbers in *italics* refer to illustrations